Alexander Lüdeke

German Half-Tracks and Wheeled Vehicles
1939–1945

German Half-Tracks and Wheeled Vehicles

1939–1945

Alexander Lüdeke

Pen & Sword
MILITARY

First published in Great Britain in 2015 by
Pen & Sword Military
an imprint of
Pen & Sword Books Ltd
47 Church Street
Barnsley
South Yorkshire
S70 2AS

ISBN 978 1 47382 400 3

Typeset in Ehrhardt by
Mac Style Ltd, Bridlington, East Yorkshire
Printed and bound in Malta by Gutenberg Press

Pen & Sword Books Ltd incorporates the imprints of Pen & Sword Archaeology, Atlas, Aviation, Battleground, Discovery, Family History, History, Maritime, Military, Naval, Politics, Railways, Select, Transport, True Crime, and Fiction, Frontline Books, Leo Cooper, Praetorian Press, Seaforth Publishing and Wharncliffe.

For a complete list of Pen & Sword titles please contact
PEN & SWORD BOOKS LIMITED
47 Church Street, Barnsley, South Yorkshire, S70 2AS, England
E-mail: enquiries@pen-and-sword.co.uk
Website: www.pen-and-sword.co.uk

Photo credits: All photos, unless another source is stated expressly, originate either from the Deutsches Wehrkundarchiv or the Alex Lüdeke Archive.

Contents

This second volume of the Typenkompass *Panzer der Wehrmacht* has as its centrepoint the armoured wheeled- and half-track vehicles used by the German forces between 1939 and 1945. Before and during the First World War the development of this kind of armoured vehicle in Germany had been undertaken in a very indolent manner. The reason for this was that Army High Command in general had little interest in armoured vehicles of whatever kind until Allied tanks appeared. Very probably the war economy, strained to breaking point, would not have been able to afford mass-produced fighting vehicles of this kind. Simply put, there was a lack of raw materials, design and completion capacity.

After 1919 the Treaty of Versailles forbade outright the development and possession of armoured vehicles.

The emphasis of the Wehrmacht panzer divisions coming into being after 1933 was naturally on fighting panzers to provide the hitting power of the Panzer Arm. Equally, assault guns, initially foreseen only for the infantry support role, and a large proportion of the SP-guns built during the war, were based on the full-track bogies of panzers.

Wheeled and half-track vehicles were however indispensable for numerous important assignments. Therefore between 1933 and 1945 an astounding multitude of types and versions of these vehicles appeared. The small numbers of armoured vehicles built in the early 1920s and permitted under the Versailles Treaty for the Reichswehr and police were already obsolete by the time they were commissioned and had very limited fighting worth. Not until the end of that decade did the Reichswehr begin to develop in secret quality armoured wheeled vehicles with tactical efficiency for reconnaissance.

For reasons of cost the development of these vehicles had to be abandoned in 1930. In their place armoured vehicles were built based on commercial lorry and saloon car bogies (Sd.Kfz.231 6-wheeler and Kfz.13/14). Neither the Reichswehr nor the Wehrmacht were under any illusions as to the usefulness of these vehicles, and orders soon went out for the development of their successors. What followed were models very efficient for the time and

Rear view of an Sd.Kfz.222, Poland, September 1939 (WKA)

they belonged without doubt to the best of their class, but the series of light and heavy armoured scout cars (Sd.Kfz.221/222/223/260/261) and the eight-wheel versions of the Sd.Kfz.231/232/233/263) were complicated in structure and maintenance. In peacetime conditions, and in the first two years of the war, they proved themselves, but in North Africa displayed their drawbacks only too clearly. The extremely harsh conditions of the campaign in Russia proved finally that the light and heavy armoured scout vehicles of the Wehrmacht were only serviceable to a limited extent.

Although a successor to the heavy scout cars of the GS type was developed and series-produced, the design for a new light armoured scout car only reached the prototype stage. The war dictated that two differing series models which basically served the same purpose were a luxury.

In order to effectively accompany and support deep and fast panzer thrusts, panzer-grenadiers had to be given the possibility of keeping up with the panzers. The solution came in the shape of light and medium armoured personnel-carriers (APC's) based on the bogie of half track towing vehicles already in Wehrmacht service. With this concept the

Wehrmacht created a completely new kind of fighting vehicle and revolutionized tank warfare.

Yet, as with the light and heavy armoured reconnaissance vehicles, in the course of the war armoured half-tracks became too expensive and complicated. The APC's from the end of 1943 were therefore greatly simplified. The difficult cross-country circumstances on the Eastern Front led to APC's taking over the armoured reconnaissance role step by step. Additionally, half-track vehicles were pressed into service as anti-tank and anti-aircraft self-propelled guns for the lack of full-track vehicles. Accordingly a multitude of light and heavy APC variants came into being, the latter becoming the numerically most proliferous Wehrmacht armoured vehicle.

All the same, the construction of two different APC's whose purposes frequently intersected or were identical was a luxury which cost valuable production capacity. Because there were never enough armoured vehicles to go round, obsolete types had to be re-used and subjected to improvisations in order to satisfy the requirements of the fighting man, if only roughly. The Wehrmacht also used practically any reasonable captured vehicle for its purposes and created, especially in occupied France, a series of its own models on the basis of these.

The enormous variety of versions resulting from this practice unfortunately render it impossible to include them all in this volume. Nevertheless I believe that this second volume in the Typenkompass *Panzers of the Wehrmacht* series, especially when combined with the first volume, provides a good overall view of the armoured vehicles used by the Wehrmacht between 1933 and 1945.

A wrecked Sd.Kfz. 251/1 version D of 2.SS.Panzer Div. Das Reich, Mortain area, France, 12 August 1944 (NARA).

Acknowledgements

My especial thanks go to my life partner Martina Pohl, whose patience made the creation of these two volumes possible, and Vincent Bourguignon, without whose sketches and fast working I could not have produced these books. Also to all those who gave me permission to use their photographic material I offer my hearfelt thanks.

I should also not fail to mention my gratitude to Motorbuch publishing house for their readiness to publish the second book upon seeing the quantity of material available.

Alexander Lüdeke
Dortmund, autumn 2008.

Glossary	
(e)	Kennz. für britische Beutefahrzeuge
(f)	Kennz. für französische Beutefahrzeuge
(h)	Kennz. für niederländische Beutefahrzeuge
(i)	Kennz. für italienische Beutefahrzeuge
(r)	Kennz. für sowjetische Beutefahrzeuge
(t)	Kennz. für tschechoslowakische Beutefahrz.
Ausf.	Ausführung
FH	Feldhaubitze
Flak	Flugabwehrkanone
FuG	Funkgerät
FuPzWg	Funkpanzerwagen
gel.	geländegängig
gep.	gepanzert
IG	Infanteriegeschütz
KwK	Kampfwagenkanone
leFH	leichte Feldhaubitze
leIG	leichtes Infanteriegeschütz
leWS	leichter Wehrmachtsschlepper
MG	Maschinengewehr
MP	Maschinenpistole
Pak	Panzerabwehrkanone
PzB	Panzerbüchse
PiPzWg	Pionierpanzerwagen
PzBefWg.	Panzerbefehlswagen
PzFuWg	Panzerfunkwagen
PzKpfw	Panzerkampfwagen
PzSpWg	Panzerspähwagen
PzWg	Panzerwagen
Sd.Kfz	Sonder-Kraftfahrzeug
Sf	Selbstfahrlafette
sFH	schwere Feldhaubitze
sGrw	schwerer Granatwerfer
sIG	schweres Infanteriegeschütz
SPW	Schützenpanzerwagen
StuG	Sturmgeschütz
StuK	Sturmkanone
sWS	schwerer Wehrmachtsschlepper
ZgKw	Zugkraftwagen

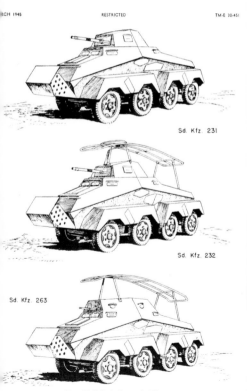

Figure 151.—Eight-wheeled armored vehicles.

Three versions of the GS-series of eight-wheeled armoured vehicles from a US Army handbook. (US Army)

Gepanzerter Mannschafts- transportwagen Sd.Kfz.3

The Treaty of Versailles signed in June 1919 prohibited the Reichswehr from possessing armoured vehicles irrespective of whether they had tracks or wheels. In 1920 the Boulogne Supplement permitted the Reichswehr to have 105 armoured troop carrier vehicles but without weapons or revolving turret. Therefore between 1921 and 1922 the firm of Daimler produced 105 vehicles known as Sd.Kfz 3 with four-wheel drive and a tall, armoured box-like superstructure. The vehicles had very narrow all-rubber tyres which rendered them only suitable for asphalted surfaces. Some were equipped with radio. Between 1921 and 1923, 85 very similar vehicles were built for the Reich police: these had two revolving turrets each equipped with a water-cooled 7.92 mm MG 08/15.

The Sd.Kfz.3 had no fixed inbuilt armament. The Reichswhr built only 105 of these vehicles.

Rear view of the Sd.Kfz.3.

In all Daimler built 31, Benz 26 (the two firms merged in 1926) and Erhardt 30 of these "Regular Police Special Vehicles" as they were termed. Depending on the manufacturer there were differences between the various chassis and in minor details. All had four-wheel drive and reverse gear, but this latter was ordered removed by the Allied Control Commission. Both the Sd.Kfz 3 and the Regular Police Special Vehicles were absorbed into the Wehrmacht in 1935 for training purposes or as radio vehicles. For the latter role some had radios fitted and the frame aerial typical of the early Wehrmacht radio vehicles.

Type:	Armoured personnel-transport car
Manufacturer:	Daimler
Fighting weight:	12 tonnes
Length:	5950 mm
Breadth:	2200 mm
Height:	2725 mm
Motor:	Daimler M1574, 4-cylinder petrol engine
Cubic capacity:	12,020 ccm
Performance kW/hp:	74/100
Performance weight:	8.3 hp/t
Top speed:	50 kms/hr (road)
Fuel capacity:	250 litres
Range:	300 km (road)
Crew:	3 + 12
Armament:	None
Armour:	7.5–10 mm
Fording depth:	0.7 m

Schupo (police) special vehicle on the DZVR bogie, seen at the Panzer Museum, Munster. (Ikeda Shinobu).

MG-Kraftwagen Kfz.13 MG-Funk-Kraftwagen Kfz.14

Some Kfz.13's even took part in the attack on the Soviet Union in mid-1941. (WKA)

Both the Kfz.13 and Kfz.14 were developed as scout cars for the Reichswehr from 1932 based on the chassis of the Standard 6, a civilian saloon car with four-wheel drive built at Adler Werke. An angular open-top armoured superstructure produced by Deutsche Edelstahl AG, Hannover-Linden was simply added to the Adler design. By 1935 Daimler-Benz had turned out 147 Kfz.13 and 40 Kfz.14.

The two-man Kfz.13 had a single 7.92-mm MG 13 with a limited field of fire and protective shield. The vehicle was not equipped with radio.

The three-man Kfz.14 was not armed and was easily distinguished from the Kfz.13 by a demountable frame-aerial for its radio installation mounted on the bodywork.

Despite their four-wheel drive these armoured cars were not good vehicles cross-country: The 8-mm armour offered poor protection even against hand-held weapons. Although both types were obsolete by 1939, they were retained in service particularly with the reconnaissance detachments of the infantry divisions and were used in the early phases of the Second World War in Poland and on the Western Front. Some even took part in the invasion of the Soviet Union in June 1941, but all were then withdrawn from active service and placed at the disposal of training units where they were gradually absorbed.

Both models were extremely useful as training vehicles but totally unsuited for the front. Only the pitiful armaments situation of the Wehrmacht made their continued use necessary into 1941. For training purposes, an imitation-armour mock-up was fitted to the Standard 6 Adler-bogie.

Type:	Light armoured scout car
Manufacturer:	Daimler-Benz
Fighting weight:	2100 kgs
Length:	4200 mm
Breadth:	1700 mm
Height:	1460 mm (Kfz.13)
Motor:	Adler Standard, 6-cylinder petrol engine
Cubic capacity:	2916 ccm
Performance kW/hp:	44/60
Performance weight:	28.5 hp/t
Top speed:	60 km/hr (road), 25 km/hr (terrain)
Fuel capacity:	70 litres
Range:	320 km (road)
Crew:	2 (Kfz.13), 3 (Kfz.14)
Armamdent:	1 x 7.92 mm MG 13 or MG 34
Armour:	5–8 mm
Fording depth:	0.5 m

MG-car (Kfz.13), 30.Inf.Div., Poland, September 1939. (WKA)

Leichter Panzerspähwagen (MG) Sd.Kfz.221

As a successor to the Kfz.13 and 14 the Wehrmacht used a series of light scout cars based on the Auto-Union bogie design I for heavy saloon cars (with engine at the rear). These bogies were all built at the Auto-Union Horch works, Zwickau, and carried the designation Horch 801. From 1935 to 1940 the version 801/EG I was built at Zwickau with a 3.5 l V8 75 hp petrol engine and mechanical braking. This bogie also had all-wheel drive, single wheel suspension, trip gear and pre-select four-wheel steering assembly and thus all variants of this light armoured scout car much more suitable for cross-country work than the Kfz.13 and 14. However the one-design bogie proved costly in maintenance prone to breakdowns. It was also complicated and expensive to manufacture.

The first such model to reach the units arrived in 1936. It was a clear improvement on its predecessors but had no radio installation. The angular, completely welded bodywork slanted to all sides was developed by Eisenwerke Weserhütte of Bad Oeynhausen, which also handled the final assembly of the Sd.Kfz.221. Armour was 14.5 mm forward, 5 to 8 mm elsewhere. Lateral hatches gave access into the interior. The two crew consisted of driver and gunner/commander in the turret. A reserve wheel was located on the bodywork right side and in the corresponding position left side a box for stowage. In the open top, hand-operated heptagonal turret was a single 7.92 mm MG 34 for which 1100 rounds ammunition would be carried. The interior also had stowage room for an MPi and ammuniton. Hinged

Sd.Kfz.221-on this light armoured (MG) scout car the rails which protected the turret against hand grenades is absent. (WKA)

Rear view of an Sd.Kfz.221 during pre-war manoeuvres. In contrast to the Sd.Kfz.222, the 221 had no seat at the turret rear.

protection against hand-grenades was provided over the open top hatch, From September 1939 in addition to the MG 34, occasionally a 7.92 mm Panzer-büchse 39 (PzB 39) anti-tank gun was added: 339 Sd.Kfz. 221 had been built by Weserhütte by 1940, all on the 801/EG I bogie. The Chinese Nationalist Government under Chiang Kai Shek bought 18 of these vehicles in 1938, the only export client for the model.

Type:	Light armoured scout car
Manufacturer:	Final assembly at Weserhütte, Bad Oeynhausen
Fighting weight:	4 tonnes
Length:	4800 mm
Breadth:	1950 mm
Height:	1800 mm
Motor:	Horch/Auto-Union, 8- cylinder V-petrol engine
Cubic capacity:	3517 ccm
Performance	kW/hp 55/75
Performance	Weight: 18.75 hp/t
Top speed:	80 km/h
Fuel capacity:	110 litres
Range:	320 km (road) 200 km (terrain)
Crew:	2
Armament:	1 x 7.92 mm MG 34
Armour:	5 to 14.5 mm
Fording depth:	0.6 m

Leichter Panzerspähwagen (2.8-cm) Sd.Kfz.221

Type:	Light armoured scout car
Manufacturer:	Final assembly at Weserhütte
Fighting weight:	4100 kg
Length:	4800 mm
Breadth:	1950 mm
Height:	1980 mm
Motor:	Horch/Auto-Union, 8-cylinder V-petrol engine
Cubic capacity:	3517 ccm
Performance:	kW/hp: 55/75
Performance weight:	18.75 hp/t
Top speed:	80 km/hr
Fuel capacity:	110 litres
Range:	300 km (road) 200 km (terrain)
Crew:	2
Armament:	1 x 2.8 cm sPzB 41 anti-tank gun
Armour:	5 to 14.5 mm
Fording depth:	0.6 m

Although the Sd.Kfz.221 had proven itself by its mobility in the opening phases of the war, the single 7.92 mm MG armament was recognized as too weak. On reconnaissance missions it was seen as essential that these vehicles were able to engage enemy armoured scout cars successfully. A PzB 39 did not improve the situation much and therefore from 1941 the Army Weapons Department ordered that with immediate effect all Sd.Kfz.221 were to be re-armed with the heavy anti-tank gun sPzB 41. For this purpose the turret front was extended and the sPzB 41 fitted with unmodified splinter shield.

The sPzB41 fired 2.8-cm shells compressed to 2-cm through a conical barrel. In order for the principle to be effective, the shell had to have a hard tungsten-carbide core surrounded by soft metal. This gave the weapon a very high muzzle velocity of

1400 m/sec., enabling the shell to pass through 66 mm armour at 500 metres (impacting surface at 90°). This was an astonishing advance but bought at the price of high wear and tear on the barrel. Tungsten was also very scarce and the ammunition hard to come by.

Sd.Kf.221 with sPzB 41, attached to 23.Pz.Div., Eastern Front, 1942. (WKA)

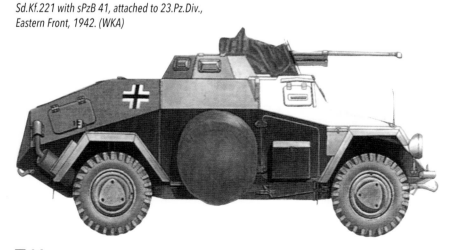

Leichter Panzerspähwagen (2 cm) Sd.Kfz.222

In 1938 an improved version with stronger armament, Sd.Kfz.222, followed the Sd.Kfz.221. This scout car was also based on the Auto-Union 801 bogie. The superstructure bore a strong resemblance to that of the Sd.Kfz. 221 designed by the Weserhütte firm, but now had a step or seat position behind the crew room. The vehicles had a new and larger ten-sided turret with a 2-cm KwK 30 or 38 and a co-axial 7.92-mm MG 34. The KwK was provided with 18 ten-round magazines, the MG had 1050 rounds available. The weaponry could be elevated to +80° for use as flak.

As with the Sd.Kfz.221 the open top hatch could be protected by a folding frame to prevent the entry of hand grenades into the interior. Crew was three, the commander, gunner and driver.

From 1938 to 1940 the light armoured scout car (2-cm) was built around the Horch 801/EG I bogie

The Sd.Kfz.222 was the Wehrmacht's standard light armoured scout car in the early years of the war. (WKA)

with 3.5 l V-petrol engine and mechanical brakes as version "A". The "B" version was completed from mid-1941 using the stronger 810/v bogie with a 90 hp 3.8 l V8-petrol engine, hydraulic brakes and other improvements. This version also had 30 mm front armour. As with the Sd.Kfz.221 the bogie was manufactured at the Auto-Union Horch works at Zwickau, the armoured upperworks came from the Eisenwerke Weserhütte firm at Bad Oeynhausen.

Final assembly was handled by the engineering firm MNH (Maschinenfabrik Niedersachsen-Hannover) and Schichau/Elbing, and later Büssing NAG.

In 1938 12 Sd.Kfz.222 were exported to China along with 18 Sd.Kfz.221. As an Axis ally of the German Reich Bulgaria also received a number of light scout cars of this type during the war.

Whereas they had proved themselves fully able to cope with the demands of combat in Poland and

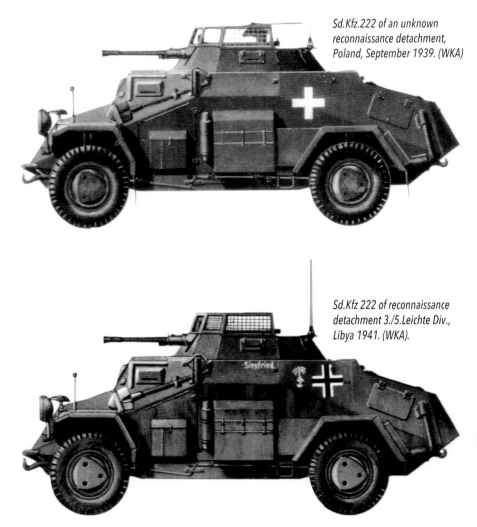

Sd.Kfz.222 of an unknown reconnaissance detachment, Poland, September 1939. (WKA)

Sd.Kfz 222 of reconnaissance detachment 3./5.Leichte Div., Libya 1941. (WKA).

Western Europe, the conditions in North Africa and especially in the Soviet Union showed the limitations of the design. Thus the ground pressure imparted through the very narrow tyres was too high, and the clearance above the ground surface of the bogie too small. In summary, the vehicles lacked cross-country worthiness and especially in snow and mud could not fulfil the task for which they had been built. Reports from the front also often mentioned damage to couplings, broken cardan shafts and springs and the sprung retaining bolts on the rocker arms of the bogie frames. The Sd.Kfz.222 was therefore classified as being of only limited battleworthiness on the Eastern Front. Although production of the type was terminated at the end of 1942 with the 989th vehicle, Sd.Kfz.222 scout cars remained in service until the end of the war.

Type:	Light armoured scout car (details vary between different types of bogie)
Manufacterer:	Final assembly at MNH, Schichau, Büssing-NAG
Fighting weight:	4800 kg
Length:	4800 mm
Breadth:	1950 mm
Height:	2 metres
Motor:	Horch/Auto-Union, 8-cylinder V-petrol engine
Cubic capacity:	3517 and 3823 ccm respectively
Performance kW/hp:	55/75 and 66/90 resp.
Performance weight:	15.6 and 18.75 hp/t resp.
Top speed:	80 and 90 km/hr resp.
Fuel capacity:	100 litres
Range:	300 km (road) 200 km (terrain)
Crew:	3
Armament:	1 x 2 cm KwK 30/38 L/55, and 1 x 7.92 mm MG 34
Armour:	5 mm to 14.5 mm (later up to 30 mm)
Fording depth:	0.6 m

Light PzSpWg (2-cm) of 16.Pz.Div., Eastern Front 1941. Notice the elevation of the 2-cm KwK and the open turret protection. (WKA)

Leichter Panzerspähwagen (Fu) Sd.Kfz.223

Type:	Light armoured scout car (radio)
Manufacturer:	Final assembly MNH
Fighting weight:	4400 and 4475 resp.
Length:	4800 mm
Breadth:	1950 mm
Height:	1750 mm (without aerial)
Motor:	Horch/Auto-Union 8-cylinder V-petrol engine
Cubic capacity:	3517 and 3823ccm resp.
Performance kW/hp:	55/75 and 66/90 resp.
Performance weight:	17 and 20.1 hp/t
Top speed:	80 to 90 km/hr
Fuel capacity:	100 litres
Range:	300 km (road), 200 km (terrain)
Crew:	3
Armament:	5 to 14.5 mm (later up to 30 mm)
Fording depth:	0.6 m

The Sd.Kfz.223 was introduced from 1938 and replaced the Kfz.14. It was basically a variant of the Sd.Kfz.222 fitted with a radio installation. Bogie, bodywork and engine therefore corresponded to the 1940 version of the 222. Armament was a 7.92 mm MG 34 with 1100 rounds carried aboard. The turret resembled that of the Sd.Kfz.221 but was nonagonal. In addition the crew had the standard issue of a machine-pistol.

After 1940 Sd.Kfz.222 version B became the basis for the 223. The weight of the vehicle was increased to 4475 kg, and a radio-operator became the third crew member. The radio installation was an FuG 10 SE 30 with a frame aerial which could be lowered if required. This aerial was very conspicuous and during the war proved so cumbersome that on later models it was replaced by a collapsible 2-metre high rod aerial, while the FuG 10 was exchanged for an FuG 12 SE 80.

The bogie was manufactured at Auto-Union and the upperworks at Weserhütte, the light radio car being assembled by MNH of Hannover. By the end of 1943 a total of 550 had come off the production line and remained in service on all fronts until the war's end. As with gthe Sd.Kfz.222 a number of 223's went to the Bulgarian ally during the war.

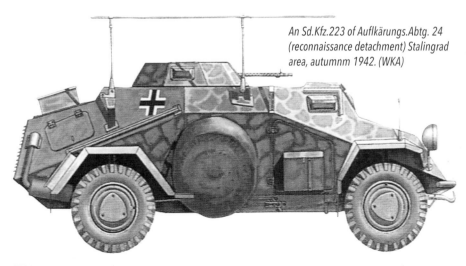

An Sd.Kfz.223 of Auflkärungs.Abtg. 24 (reconnaissance detachment) Stalingrad area, autumnm 1942. (WKA)

An Sd.Kfz.223 followed by an Sd.Kfz.221 in the Russian winter. (WKA)

Sd.Kfz.223, 21.Panzer-Div.Gazala, 1941. (WKA)

Sd.Kfz.260 (NARA)

Leichter Panzerfunkwagen Sd.Kfz. 260

the four-man crew, an MPi and six magazines (192 rounds) was stowed in the interior.

By 1943 a total of 493 Sd.Kfz. 260 and 261 had been produced, the former at MNH and the latter at

Both versions of this light armoured radio car served primarily as armoured radio centres for troop and unit commanders to contact the regimental or divisional HQs. Similar to the Sd.Kfz.221, 222 or 223, they were based on the universal bogie I frame of the Auto-Union heavy car (with rear drive). The principal difference between the 260 and 261 models was the version of the bogie and the radio installation. The Sd.Kfz.260 had the Type 801/EG I (3.5 litre 75-hp engine) bogie and a medium range radio with rod aerial. The Sd.Kfz.261 was based on the improved Type 801/v (3.8 litre 90-hp engine), its radio having greater range, using a frame aerial to send and receive. Later this was replaced by a rod aerial. Both vehicles were usually not fitted with armament, besides the personal weapons issue of

An Sd.Kfz.261 of an unknown unit captured by US troops near Salerno, end of September 1943. (NARA)

ITALY

Weserhütte. Although they were gradually being replaced by similarly equipped armoured half-track vehicles, armoured cars of this kind remained in use until the end of the war.

An Sd.Kfz.261 with frame aerial. (Vincent Bourguignon).

Type:	Light armoured radio car	
	Sd.Kfz. 260	Sd.Kfz. 261
Manufacturer:	Final assembly MNH	Final assembly Weserhütte
Fighting weight:	4260 kg	4300 kg
Length:	4830 mm	4830 mm
Breadth:	1990 mm	1990 mm
Height:	1780 mm	1780 mm
Motor:	Horch/Auto-Union 8-cylinder V-petrol engine	Horch/Auto-Union 8-cylinder V-petrol engine
Cubic capacity:	3517 ccm	3823 ccm
Performance kW/hp:	55/75	66/90
Performance weight:	17.6 hp/t	20.9 hp/t
Top speed:	80 km/hr	90 km/hr
Fuel capacity:	110 litres	100 litres
Range:	320 km (road) 200 km (terrain)	300 km (road) 200 km (terrain)
Crew:	4	4
Armament:	None	None
Armour:	5 to 14.5 mm (later 30 mm)	5 to 30 mm
Fording depth:	0.6 m	0.6 m

Schwerer geländegängiger gepanzerter Personenkraftwagen Sd.Kfz. 247 Ausf.B (4-wheeler)

Type:	Armoured car
Manufacturer:	Daimler-Benz
Fighting weight:	4460 kg
Length:	5 m
Breadth:	2 m
Height:	1800 mm
Motor:	Horch/Auto-Union 8-cylinder V-petrol engine
Cubic capacity:	3823 ccm
Performance kW/hp:	60/81
Performance weight:	18 hp/t
Top speed:	80 km/hr
Fuel capacity:	120 litres
Range:	400 km (road) 270 km (terrain)
Crew:	6
Armament:	None
Armour:	5 to 8 mm
Fording depth:	0.6 m

Between 1938 and January 1942 Daimler-Benz built 58 lightly armoured Staff vehicles at its Marienfelde factory in Berlin. Like the A six-wheel version of the Sd.Kfz.247 the purpose was to enable commanders to be at the frontline. The B version also had the common Auto-Union bogie, although variant II had the engine and gearing located at the front. The open armoured superstructure was of welded steel plates of an 8 mm maximum thickness. The Sd.Kfz.247 version B was not armed but carried an MPi and 192 rounds in six magazines in the interior. Although designed primarily as a command vehicle, it was not originally equipped with radio. This was added sometime later with a very characteristic cluster of aerials, and in this configuration the cars were used mainly as armoured radio centres. Sd.Kfz.247 version B were operated almost exclusively by Waffen-SS units and the elite Grossdeutschland Division.

An Sd.Kfz.247 of Panzergrenadier-Division Grossdeutschland, 1942. (Vincent Bourguignon)

Leichter Panzerspähwagen Tp4 (Prototyp)

Type:	Light armoured scout car
Manufacturer:	Büssing-NAG
Fighting weight:	c. 7 tonnes
Length:	5330 mm
Breadth:	Not available
Height:	2150 mm
Motor:	Tatra, air-cooled 6-cylinder diesel engine
Cubic capacity:	Not known
Performance kW/hp:	78/125
Performance weight:	17.6 hp/t
Top speed:	85 km/hr (road)
Fuel capacity and range:	Not known
Crew:	4
Armament:	1 x 2 cm KwK 38, 1 x 7.92 mm MG 42
Armour: max.	30 mm
Fording depth:	Not known

As a successor to the range of light armoured scout cars Sd.Kfz.221/222/223, the Army Weapons Department ordered a vehicle with as many components as possible to be the new eight-wheeler scout car "Tp".

On 21 July 1941 Büssing-NAG received the contract to design and build the prototype. Apart from the engine, transmission and suspènsion and a modified rear to the armoured superstructure the design corresponded as closely as possible to that of the Büssing-NAG Sd.Kfz.234. Analogous to the 234 two different armed versions were planned. The first would have a hexagonal turret open at the top equipped with a 2-cm KwK 38 and co-axial MG 42 with great elevation for use against aircraft. The other had a turret with a 5-cm KwK 39/1 and co-axial MG 42. Büssing-NAG completed two prototypes by the spring of 1943. Although a contract was issued for 1000, production on which was scheduled to begin from the autumn of 1943, the vehicle never entered series production.

Based on components of the Sd.Kfz.234, this prototype was to have been the successor to the Sd.Kfz.222.
(Vincent Bourguignon)

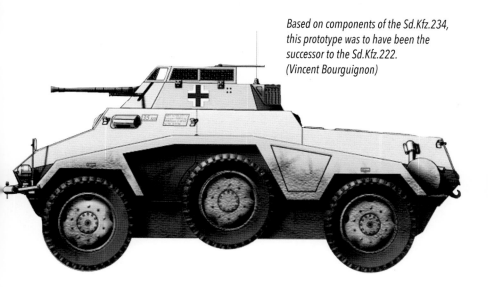

Mittlerer gepanzerter Beobachtungs- kraftwagen Sd.Kfz.254

Type:	Medium armoured observation vehicle
Manufacturer:	Saurer
Fighting weight:	6400 kg
Length:	4500 mm
Breadth:	2400 mm
Height:	2 metres on tracks, 2.2 metres on wheels
Motor:	Saurer CRDv 4-cylinder diesel engine
Cubic capacity:	5300 ccm
Performance kW/hp:	51.5/70
Performance weight:	10.9 hp/t
Top speed:	60 km/hr (wheels), 30 km/hr (tracks)
Fuel capacity:	72 litres
Range:	425 km (road) 90 km (terrain)
Crew:	4
Armament:	1 x 7.92 mm MG 34
Armour: max.	15 mm
Fording depth:	0.65 m

The Sd.Kfz.254 was an unusual design being a so-called wheeled-tracked vehicle able to exchange one form of traction for the other during travel. The origin of the vehicle was the unarmoured tug Saurer RR-7 of which the Austrian Federal Army had ordered fifteen before the annexation of Austria in 1938. From May 1938 the Weapons Department commissioned from Saurer an improved variant of the RR-7 for the Wehrmacht, the prototype for which was not forthcoming until June 1942. This first prototype called RK-9 Ausf. (version) A was not armed, a second model received a turret for an MG and anti-tank gun developed by Daimler-Benz. This three-man vehicle had armour 5.5 to 14.5 mm thick,

drive was provided by a 100-hp Saurer diesel engine. Although a 0-series of fifteen was ordered, this further development did not enter production. Instead Saurer had built for the Wehrmacht by the end of 1940 a total of 140 practically unmodified

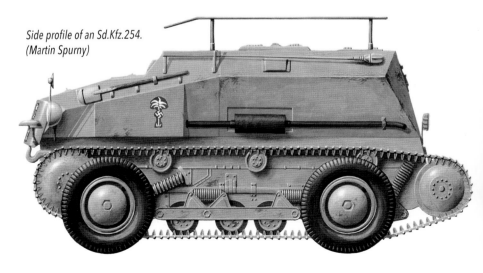

Side profile of an Sd.Kfz.254. (Martin Spurny)

An Afrika Korps Sd.Kfz.254 with frame aerial. (WKA)

RR-7 though with an armoured superstructure open at the top. Apparently only 128 of these were accepted by the Army Weapons Department.

The type had no fixed weaponry but carried along a 7.92 mm MG 34. They were used primarily as armoured scout cars and as artillery spotters. Some Sd.Kfz.254 had radio installations and the corresponding frame aerials. The RR-7 failed to live up to the expectations of military routine. An unarmoured version of the RR-7 was used by the Wehrmacht as a repairs waggon with a 1.5 tonne payload.

The 140 examples of this wheeled/tracked vehicle failed to make their mark, being too prone to breakdown mechanically. (WKA)

Panzerspähwagen PA-II (t)

Type:	Armoured scout car
Manufacturer:	Skoda
Fighting weight:	7360 kg
Length:	6 m
Breadth:	2160 mm
Height:	2440 mm
Motor:	Skoda 4-cylinder petrol engine
Cubic capacity:	9730 ccm
Performance kW/hp:	51/70
Performance weight:	9.5 hp/t
Top speed:	70 km/hr (road)
Fuel capacity:	175 litres
Range:	250 km (road)
Crew:	4 to 5
Armament:	4 x 7.92 mm MG Maxim 08
Armour:	3 to 5.5 mm
Fording depth:	0.4 m

In 1924 and 1925 Skoda built a total of twelve Type PA-II (unofficially also designated OA vz.23) armoured cars with four wheel drive which could be driven as facing forward either way. The rounded, symmetrical armoured superstructure without turret gave the vehicle its nickname "Zelva" (tortoise). Armament was four MG 7.92 mm calibre MGs with 6,250 rounds built into bullet shields on four sides of the superstructure. The off-road performance of the vehicles was very limited, as a result of which the Czech Army passed them to the State police after the first trials.

After Czechoslovakia was occupied by Germany in March 1939, the remaining nine PA-II joined the German stocks of fighting vehicles. Nothing is known about the purpose to which they were put although there is a photo from the outset of the Russian campaign which shows a PA-II in Wehrmacht service

unarmed and equipped with a frame aerial. Three PA-II sold to the Vienna police force in 1927 were allegedly used in the defence of the Austrian capital in 1945.

Skoda PA-II in Czech Army livery.
(Martin Spurny)

Panzerspähwagen 27 (t)

Type:	Armoured scout car
Manufacturer:	Skoda
Fighting weight:	6600 kg
Length:	5350 mm
Breadth:	1950 mm
Height:	2660 mm
Motor:	Skoda 4-cylinder petrol engine
Cubic capacity:	5700 ccm
Performance kW/hp:	51/60
Performance weight:	9.5 hp/t
Top speed:	35 km/hr (road) 20 km/hr (terain)
Fuel capacity:	Not known
Range:	250 km/hr (road)
Crew:	5
Armament:	2 x 7.92 mm MG vz.7/24, 1 x 7.92 mm MG vz.26
Armour:	3 to 5.5 mm
Fording depth:	Not known

As a result of the deficiencies of the PA-II, in 1927 Skoda developed a successor designated PA-III which was introduced into the Czech Army as OA vz.27 in 1929. This bizarre-looking vehicle had four-wheel drive, could be driven as facing forward either way and had bullet-proof tyres.

A 7.92-mm MG vz.7/24 was mounted in the turret, another MG of the same type was located in a bullet shield between driver and co-driver. An armoured searchlight was built-in at the rear of the turret. A light Type vz.26 MG and a total of 5750 rounds for all three MGs was carried inside the vehicle. From 1933 the turret-MG on some of the vehicles was exchanged for a 2-cm cannon. Skoda built 16 PA-III of which three went to Rumania and three to Slovakia after Czechoslovakia was dismembered, the remainder were acquired by the Wehrmacht, little information exists as to their further use.

The Wehrmacht took over ten Skoda PA-III as armoured scout cars 27(t). In this example to turret is facing to the rear. (Martin Spurny)

Panzerspähwagen P 204(f)

The Panhard AMD-178 was designed in the mid-1930s for the French Army. Up to the time of the Western campaign in May 1940 about 480 of them had been delivered, which made Panhard the leading supplier of French armoured vehicles. The AMD-178 was modern and possessed interesting innovations, amongst them for example a second driver facing to the rear who in dangerous situations could take over the controls, and the two-stroke engine with rotary sliding valves instead of blowers. A drawback was the bolted-on armour. If the car was hit, these bolts could pop and fly around the crew area. Besides the version armed with a 2.5-cm cannon and a co-axial 7.5 mm MG, a variant had a 7.5-mm twin MG. After the defeat of France the Wehrmacht had all available AMD-178's given a general overhaul at the Panhard works in Paris and then took them into service as armoured scout cars P 204(f) equipped with a German radio installation,

some with the typical frame aerial. In action the vertical armour was found to give insufficient protection and the troops often mounted additional armour at the endangered spots.

It is not known exactly how many P 204(f) were in German service, but it is estimated to be over 200.

Initially the P 204(f) was used by reconnaissance units, but later more commonly by security and police units. In 1942, 43 vehicles were converted to run on railway metals. In this form they were used as security for armoured railway trains or to control and protect stretches of track against partisans, especially in the East and Balkans.

From June 1940 there were still around 40 to 50 Panhards in the unoccupied part of France and used there by the Vichy forces. After the occupation of Vichy France in November 1942 most of these

Armoured scout car 204(f), unit unknown, Eastern Front 1943. (Vincent Bourguignon)

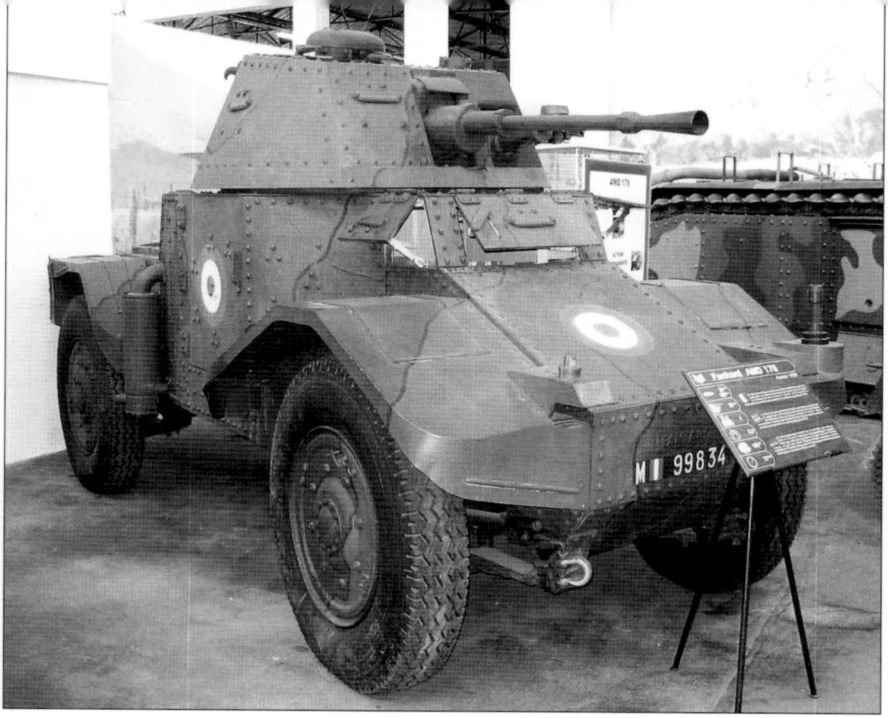
M 99834
Panhard AMD 178

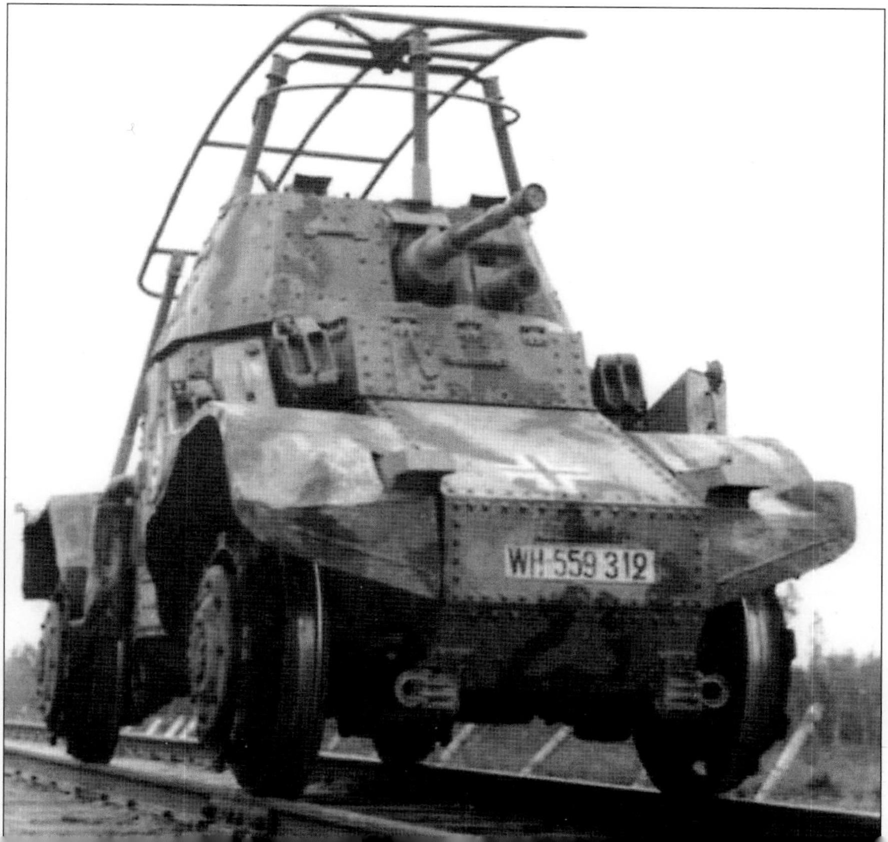
WH 559 312

vehicles then fell into German hands. The turret was removed from some of them and replaced by a fixed, open superstructure in which a 5-cm KwK L/42 was fitted. Those Panhards captured in November 1942 were used mainly on the Western Front.

Up to their defeat in June 1940, French forces used 24 unarmed radio cars as command vehicles. This variant of the Panhard armoured scout car had the revolving turret replaced by a fixed superstructure. High rod aerials were fitted on the mudguards left side front and rear. Some of these cars fell into German hands, were given a 100 Watt radio installation FuG 11 and designated "armoured radio post Panhard".

A number of these vehicles were used by war correspondents and propaganda companies and fitted with special colour photography equipment and cameras. An MG 34 was integrated into the front of the car for self-defence. The Wehrmacht was very keen on the Panhard and considered it a valuable addition to their reconnaissance units.

Type:	Armoured scout car
Manufacturer:	Panhard, France
Fighting weight:	8300 kg
Length:	5140 mm
Breadth:	2010 mm
Height:	2330 mm
Motor:	Panhard SS 4-cylinder petrol engine
Cubic capacity:	6330 ccm
Performance kW/hp:	78/105
Performance weight:	15.1 hp/t
Top speed:	72 km/hr (road)
Fuel capacity:	125 litres
Range:	350 km (road 130 km (terrain)
Crew:	4
Armament:	1 x 2.5 cm KwK, 1 x 7.5 mm MG
Armour:	7 to 20 mm
Fording depth:	0.6 m

Panhard radio car. (WKA)

Max was one of two AEC armoured command vehicle used by Rommel's Staff. (WKA)

Kommando-panzerwagen AEC Matador

From 1941 the British manufacturer AEC produced 415 so-called "armoured command vehicles" on the bogie of the heavy 4 x 4 lorry *Matador*. These vehicles with a large armoured box-like bodywork and comprehensive radio installation were used principally by the British Army for senior commanders and telecommunications. They were nicknamed "Dorchester" after the famous London hotel for their spacious (for the time) and comfortable interior. The Afrika Korps captured three of them at Mechili in Libya in April 1941 calling them "Mammut" (mammoth) and kept them in service until May 1943. Two were used by Rommel and his Staff as command vehicles and were known as Max and Moritz. The third was used by General Streich.

Type:	Armoured command vehicle
Manufacturer:	AEC
Fighting weight:	12,200 kg
Length:	6340 mm
Breadth:	2360 mm
Height:	2900 mm
Motor:	AEC 187 6-cylinder diesel engine
Cubic capacity:	5700 ccm
Performance kW/hp:	71/95
Performance weight:	7.8 hp/t
Top speed:	60 km/hr (road)
Fuel capacity:	182 litres
Range:	450 km (road)
Crew:	8
Armament:	None
Armour:	10 to 12 mm
Fording depth:	Not known

Leichter Panzerspähwagen BA-20 202 (r)

Type:	Light armoured car
Manufacturer:	Soviet State
Fighting weight:	2500 kg
Length:	4300 mm
Breadth:	1750 mm
Height:	2310 mm
Motor:	GAZ M1 4-cylinder petrol engine
Cubic capacity:	3300 ccm
Performance kW/hp:	37/50
Performance weight:	20 hp/t
Top speed:	85 km/hr (road) 28 km/hr (terrain)
Fuel capacity:	Not known
Range:	350 km (road) 200 km (terrain)
Crew:	2 to 3
Armament:	1 x 7.62 mm MG of type DT
Armour: max.	10 mm
Fording depth:	Not known

In the opening weeks and months of the attack on the Soviet Union, German forces captured large quantities of Soviet weaponry including numerous armoured scout cars such as vehicles of the type BA-20. These were based on a civilian bogie of the GAZ-M1, itself a further development of a Ford design. The BA-20 was produced from 1936 and by 1941 the Red Army had large numbers of them in service. The two-axled car had only rear-wheel drive which seriously limited its manouevrability cross-country. Besides the reconnaissance version the BA-20V was a radio car with frame aerial, and the BA-20 ZhD served as a railway trolley. From 1938 the version BA-20M was built with a modified turret, standard radio installation with whip aerial and a larger fuel tank. The Wehrmacht used captured versions as both scout cars and for security purposes.

A BA-20 ZhD with frame aerial in German service. Vehicles of this type were often used to protect stretches of railway line. (Vincent Bourguignon)

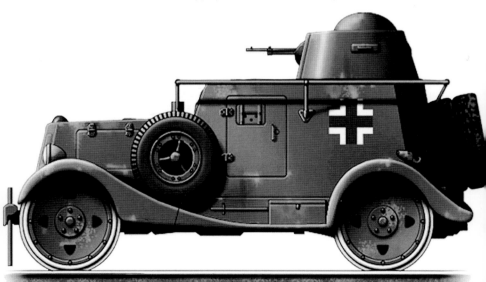

Panzerspähwagen AB 41 201 (i)

After Italy had concluded a separate armistice with the Western Allies at the beginning of September 1943, the German forces gathered up all the war material they could lay hands on. Amongst it all was a large number of armoured wheeled vehicles. Together with armoured cars of the Lancia IZM type and around three dozen armoured Carro Protetto A.S. 37 troop transport vehicles, the majority were armoured scout cars of types AB 40 and 41.

Type AB 41 was a further development of the model AB 40 used originally as an armoured car by the Italian police in the African colonies and also by cavalry units.

The AB 40, turned out from mid-1940 by SPA (a branch of the Fiat company) had all-wheel drive and controls, a steering wheel forward and rear for travel in either direction and excellent manoeuvrability cross-country. Initially only an 8 mm twin MG was fitted in the revolving turret and a single MG of the same calibre placed alongside the rear-facing driver. In order to increase the firepower, after a few models had been produced, the turret of the light tank L6/40, which had a 2-cm cannon and a co-axial MG, was fitted experimentally and the production run was

An "Autoblinda AB 41" of the Italian Army in the North African desert. (WKA)

completed with the vehicle now being called the AB 41. Additionally a more powerful engine needed to be installed in view of the gain in weight from 6.8 to 7.5 tonnes due to the new turret. Many of the earlier AB 40s were similarly re-equipped. This vehicle was very progressive and efficient for the time but plagued with steering problems never completely eradicated. Some of the cars also had a flamethrower-MG on the turret. Equipment was available to convert the vehicle to run on railway lines, use often being made of it in the Balkans. A command variant with an open-top armoured superstructure was also built in very small numbers.

In September 1943 about 200 of these vehicles were taken over by the Germans, around 110 being hived off to Croat units. Thus at the beginning of October 1943 the Wehrmacht had exactly 87 AB 40 and 41's. Since the manufacturer was located in the Wehrmacht-controlled region of Italy production for the German forces was continued there, but by the end of 1944 only another 23 armoured cars of type AB 41 had been turned out. Already before Italy left the Axis Pact, SPA had been working on the prototype of the successor designated AB 43. Along with the 2-cm weapon a 4.7-cm KwK L/32 or L/40 could be fitted in a flatter but more roomy turret if so desired. In addition as an option the lateral support/spare wheels could be dispensed with, the access hatches forward repositioned and modifications made at the rear. In October 1943 the Wehrmacht issued a contract for 310 AB 43, of which only 60 were forthcoming in 1944 and an unknown number in 1945. The vehicles turned out were basically AB 41's fitted with the new AB 43 turret.

An AB 41 201(i) deployed on the Eastern Front. (NARA)

Type:	Armoured scout car (data for AB 41)
Manufacturer:	SPA-nsaldo
Fighting weight:	7500 kg
Length:	5200 mm
Breadth:	1920 mm
Height:	2480 mm
Motor:	SPA Abm 1 6-cylinder petrol engine
Cubic capacity:	4995 ccm
Performance kW/hp:	65/88
Performance weight:	11.7 hp/t
Top speed:	78 km/hr (road) 38 km/hr (terrain)
Fuel capacity:	118 litres
Range:	300 km (road)
Crew:	4
Armament:	1 x 2 cm cannon M35, 2 x 8 mm Breda MG M 38
Armour: max.	9 mm
Fording depth:	0.7 m

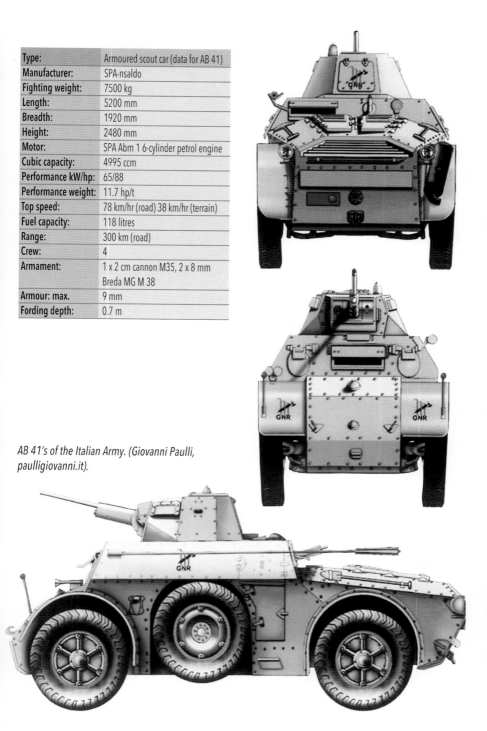

AB 41's of the Italian Army. (Giovanni Paulli, paulligiovanni.it).

Leichter Panzerspähwagen Mk I 202 (e)

Type:	Light armoured scout car
Manufacturer:	Daimler, Humber
Fighting weight:	3 tonnes
Length:	3180 mm
Breadth:	1710 mm
Height:	1500 mm
Motor:	Daimler 6-cylinder petrol engine
Cubic capacity:	2510 ccm
Performance kW/hp:	41/55
Performance weight:	18.4 hp/t
Top speed:	88 km/hr (road)
Fuel capacity:	Not known
Range:	320 km (road)
Crew:	2
Armament:	1 x 7.7 or 7.92 mm MG
Armour: max.	30 mm
Fording depth:	1 m

This light armoured scout car was designed at BSA in 1938, but production ensued the following year at Daimler. Although the official title was the Daimler Scout Car, it was much more widely known as the Dingo. It had independent wheel suspension, bullet-proof tyres, all-wheel drive and selection. By the war's end 6,626 vehicles had been manufactured at Daimler and Humber in the slightly differing model series I to III. In Canada, Ford built 3,255 examples known as the Lynx. Both models were equipped with radio and a 7.7 mm MG or a 14 mm anti-tank gun. In France, Greece and North Africa the Wehrmacht captured large numbers of the versatile Dingo and commissioned them under the designation "light armoured scout car MK I 202(e)" as liaison, reconnaissance and security vehicles. Some front units thrust them into service without any modification but gradually they were given German armaments and radio installations and incorporated into Wehrmacht units in the normal way.

PzSpWg Mk 1 202(e) were used not only by the German Army but also in motorized units of the Kriegsmarine.

Leichter Panzerspähwagen Lince 202 (i)

Type:	Light armoured scout car
Manuafacturer:	Lancia
Fighting weight:	3140 kg
Length:	3200 mm
Breadth:	1600 mm
Height:	1760 mm
Motor:	Lancia Astura 8-cylinder petrol engine
Cubic capacity:	2617 ccm
Performance kW/hp:	44/60
Performance weight:	19.1 hp/t
Top speed:	86 km/hr (road)
Fuel capacity:	Not known
Range:	300 km (road)
Crew:	2
Armament:	1 x 8 mm Breda MG
Armolur: max.	**14 mm**
Fording depth:	1 m

The Italian Army captured in North Africa a number of Daimler Dingo scout cars which impressed them sufficiently for the Italian leadership to decide upon making an almost identical copy. Production at Lancia had not started by September 1943 and not until 1944 under German supervision at Turin was series production of the all-wheel drive vehicle with independent suspension finally undertaken. The Wehrmacht gave Lancia a contract for over 300 Lince of which only 104 had reached German troop units by the end of 1944. Production continued into 1945 but the number made is not known. These scout cars formed German armoured scout car detachments, as armoured liaison vehicles or worked together with Ab 41/43 armoured cars (also of Italian origin). Armament was an 8 mm Breda MG of Italian manufacture. These vehicles were also used by Italian troops of the Repubblica Sociale Italiana, RSI fighting alongside the Germans.

The PzSpWg Lince (=lynx) was an almost servile copy of the British Dingo. (WKA)

Schwerer Panzerspähwagen Sd.Kfz. 231 (six-wheeler)

After experiments in the late 1920 with technically advanced armoured cars had had to be terminated on grounds of expense at the beginning of the 1930s, the Reichswehr ordered a rapid and economical solution to armoured car manufacture. Thus from 1930 three slightly different models of a heavy armoured scout car designated Sd.Kfz.231

Sd.Kfz 231 (six-wheeler), Poland, September 1939. (WKA)

Sd.Kfz.231 (six wheeler) in Dürnast village during the 1935 IX.Army Korps autumn manouevres in the Bavarian Ostmark. (BA cc-by-sa 3.0)

were built on merchant lorry bogies, the M206 Magirus, the Büssing-NAG G31 and the Daimler-Benz G3 (and G 3a). These 6 x 4 bogies all had water-cooled petrol engines with slightly different output (Magirus 70 hp, Büssing-NAG 60 hp and Daimler-Benz 68 hp) working on the rear axle. In order to meet the Army Weapons Department specifications

for an armoured vehicle, the axles of all three were reinforced, the cooling system improved and the possibility of rear steering introduced. The cars had reverse-drive gearing for travel in either direction. The Magirus bogie had a top speed in reverse of 100%, Daimler-Benz 75% and the Büssing-NAG version 50%. All three versions had identical armoured superstructures developed by Deutsche Werke, Kiel, completed both there and also at Deutscher Edelstahl AG, Hannover-Linden. In order to provide greater protection against incoming fire, as far as possible the armour was slanted towards the main risk directions. The front armour was 14.5 mm, 8 mm at the sides and rear. Access into the vehicle was by a rear door and two hatches between the first and second axles. Two cupola-like bulges projected above the seats of the front and rear drivers. Weaponry was a 2-cm KwK 30 L/55 and a 7.92 mm MG 13 in a hand-operated turret with hatches at the top and rear. 200 rounds of ammunition were carried for the KwK, 1300 to 1500 for the MG. It was not planned to instal a radio installation in the basic "gun car" version.

To improve traction on soft ground, rubber tracks could be fitted to the rear pair of wheels. Nevertheless because of the high pressure on the ground through the narrow tyres, the poor clearance and lack of all-wheel drive the cross-country characteristics of the vehicles were anything but adequate. Reichwehr and Wehrmacht standing orders made it very clear that vehicles of this type were not to leave paved surfaces unless unavoidable. In addition the weight of the armoured superstructure overburdened the bogie.

The first Sd.Kfz.231 were introduced into the Reichswehr in 1932 and seen on manoevres for the first time that year. Armoured cars of this type were only thought of as an interim solution before the advent of eight-wheeler cross-country types which, having the same purpose would therefore receive the same Sd.Kfz. number. Because both types were jointly in service for a considerable period, from 1937 it was common to add after the designation Sd.Kfz.231 either the suffix (six-wheeler) or (eight wheeler). So long as a good network of streets and

Although partly obscured by trees, the lateral access hatches are easily seen on this Sd.Kfz.231 (six-wheeler).

Type:	Heavy armoured scout car (data for the Büssing-NAG version)
Manufacturer:	Büssing-NAG, Magirus, Daimler-Benz
Fighting weight:	5350 kg
Length:	5570 mm
Breadth:	1820 mm
Height:	2250 mm
Motor:	Büssing-NAG G 4-cylinder petrol engine
Cubic capacity:	3920 ccm
Performance kW/hp:	44/60
Performance weight:	11.2 hp/t
Top speed:	65 km/hr (road)
Fuel capacity:	90 litres
Range:	260 km (road) 140 km (terrain)
Crew:	4
Armament:	1 x 2 cm KwK 30 L/55, 1 x 7.92 mm MG 13
Armour	8 to 14.5 mm
Fording depth:	0.6 m

roads were available the Sd.Kfz (six wheeler) was thoroughly practical and was accordingly deployed in the occupation of the Sudetenland, Czechoslovakia, Poland and in the West. After the summer of 1940 the six-wheeler was withdrawn from the fronts and used only for training or security purposes.

Sd.Kfz.231 (six wheelers) at a pre-war parade in Hitler's presence.

r view of an Sd.Kfz.231 (six
eeler) with turret facing back. The
tion from this pre-war postcard
d: "Heavy armoured scout car on
post duty".

Schwerer Panzerspähwagen (Fu) Sd.Kfz.232 (six wheeler)

Besides the basic "gun car" version of the Sd.Kfz.231 (six wheeler), a radio car variant was introduced from 1933, the Sd.Kfz.232. A characteristic of the radio car was the large aerial framework located above the superstructure which though not limiting the traverse of the turret raised appreciably the height of the vehicle. Because the 100-Watt radio installation made possible contact to rearward command positions, the Sf.Kfz.232 was used mainly by unit commanders. Apart from the radio and aerial there were no significant differences from the basic version. Armoured superstructure, rotatable turret, armament, ammunition storage and crew numbers were identical. The six-wheeler radio car had the same drawbacks as the gun car. Because of the shortage of scout cars, until mid-1940 the

As can be seen from this pre-war photo, the turret remained fully rotatable despite the aerial frame. (WKA)

Wehrmacht used the six wheeled radio car attached to reconnaissance units even though it had only been conceived as an interim solution until eight-wheeled vehicles reached the units. Information respecting the numbers of six-wheeler Sd.Kfz.231 and 232 manufactured between 1930 and 1935 is widely diverse. The literature speaks of over 900 vehicles built but realistically it can only have been 123; 36 at Daimler-Benz, 12 at Büssing-NAG and 75 at Magirus.

Sd.Kfz.232 (6-wheeler) during VI.Armee Korps manouevres on Lüneburg Heath between 2. and 7 September 1935.

Type:	Heavy armoured radio scout car (Daimler-Benz version)
Manufacturer:	Büssing-NAG, Magirus, Daimler-Benz
Fighting weight:	5700 kg
Length:	5570 mm
Breadth:	1820 mm
Height:	2870 mm
Motor:	Daimler-Benz M09 6-cylinder petrol engine
Cubic capacity:	3663 ccm
Performance kW/hp:	50/68
Performance weight:	11.9 hp/t
Top speed:	70 km/hr (road)
Fuel capacity:	105 litres
Range:	300 km (road) 150 km (terrain)
Crew:	4
Armament:	1 x 2 cm KwK 30 L/55, 1 x 7.92 mm MG 13
Armour:	8 to 14.5 mm
Fording depth:	0.6 m

Driving past the rostrum during a Nuremberg Rally pre-war. (WKA)

45 ■

Schwerer Panzerfunkwagen Sd.Kfz. 263 (six-wheeler)

Type:	Heavy armoured radio scout car (Daimler-Benz version)
Manufacturer :	Büssing-NAG, Magirus, Daimler-Benz
Fighting weight:	5800 kg
Length:	5570 mm
Breadth:	1820 mm
Height:	2870 mm
Motor:	Daimler-Benz M09 6-cylinder petrol engine
Cubic capacity:	3663 ccm
Performance kW/hp:	50/68
Performance weight:	11.7 hp/t
Top speed:	70 km/hr (road)
Fuel capacity:	105 litres
Range:	300 km (road) 150 km (terrain)
Armament:	1 x 7.92 mm MG 13
Armour:	8 to 14.5 mm
Fording depth:	0.6 m

The heavy armoured radio car (Sd.Kfz.263) bore a strong resemblance to the Sd.Kfz.232, both versions having identical armoured superstructures and aerial frameworks. In their equipment and inner arrangement however there were major differences between the two versions, for the Sd.Kfz. was primarily a command car and armoured command post for senior commanders or Staffs. Accordingly the vehicle had a more efficient radio than the Sd.Kfz.232 and a telescopic aerial which could be disassembled. The mast could be cranked up by hand to a height of 9 metres. To create room in the interior for a fifth man, map table and radio installation, the turret rotating gear and 2-cm KwK 30 were dispensed with and replaced by a single 7.92-mm MG 13 at the front of the fixed turret. After the campaign in the West, the Sd.Kfz.263 (six-wheeler) disappeared from the front and went to training units. A total of 28 heavy armoured radio cars of this type were built.

Besides a frame aerial the heavy armoured radio car had an extending aerial, here hidden below a tarpaulin. (WKA)

Schwerer gelände-gängiger gepanzerter Personenkraft-wagen Sd.Kfz. 247 Ausf.A.

Type:	Heavy cross-country armoured Staff car
Manufacturer:	Daimler-benz
Fighting weight:	5200 kg
Length:	4600 mm
Breadth:	1920 mm
Height:	1700 mm
Motor:	Krupp M304 4-cylinder petrol engine
Cubic capacity:	3308 ccm
Performance kW/hp:	38/52
Performance weight:	10 hp/t
Top speed:	70 km/hr (road)
Fuel capacity:	110 litres
Range:	350 km (road 220 km (terrain)
Crew:	6
Armament:	None
Armour:	8 mm
Fording depth:	0.6 m

In 1946 Daimler-Benz received from the Army Weapons Department a contract to develop an armoured Staff car enabling commanders to lead their troops from the front. The vehicle was also to be deployed by reconnaissance units. Daimler-Benz designed the Sd.Kfz.247 Ausf.A and built it from start to finish at their Marienfelde works in Berlin, although Krupp made the L 2 H 143 bogie, (this being also the basis of the 1.5-tonner Krupp limber or Kfz.69.) This three-axled bogie had a front-mounted Boxer engine driving the two rear axles. The open-top superstructure with all-round 8 mm armour was manufactured by Deutscher Edelstahl AG

Hannover-Linden and could be covered over by a tarpaulin if required. Apart from an MPi with 192 rounds in the interior the vehicels were not armed and had no radio installation but often carried scissor optics for observation purposes. Daimler-Benz produced only ten Sd.Kfz.247 Ausf.A up to 1937.

Sd.Kfz.247 Ausf. A of an unknown unit, Poland, September 1939. (Vincent Bourguignon)

Leichter Panzerspähwagen 30(t)

Type:	Light armoured scout car
Manufacturer:	Tatra
Fighting weight:	2780 kg
Length:	4020 mm
Breadth:	1520 mm
Height:	2020 mm
Motor: Tatra	71 4-cylinder petrol engine
Cubic capacity:	1910 ccm
Performance kW/hp:	24/32
Performance weight:	11.5 hp/t
Top speed:	60 km/hr (road)
Fuel capacity:	Not known
Range:	300 km (road)
Crew:	3 to 4
Armament:	2 x 7.92 mm MG ZB vz.26
Armour:	3 to 6 mm
Fording depth:	Not known

After the occupation and break-up of Czechoslovakia in March 1939, numerous armoured vehicles fell into Wehrmacht hands. The most important of these numerically was the Type OA vz.30, called "light armoured scout car 30(t)" in German service. The vehicle was based on the Tatra T 72 bogie with three axles, the two at the rear providing the drive. The two light MGs were carried one in the turret, the other inbuilt alongside the driver. From 1933 Tatra built a total of 51 of this type for the cavalry units of the Czechoslovak Army. After March 1939 eighteen of them went to the Slovak, nine to the Rumanian and one to the Hungarian armies. The other twenty-three were used by the German forces and police. The Wehrmacht had seven of these converted into radio broadcast reception vehicles so as to make sound recordings at the front to accompany war bulletins. Some vehicles were used as unarmed command cars and fitted with radio and frame aerials.

OA vz.30 in Czech Army livery. The MG which should be mounted at the front alongside the driver is absent. (Martin Sprurny)

Unarmed command version of the PzSpWg 30(t) with frame aerial, 1940. (Vincent Bourguignon)

Panzerspähwagen DAF 201 (h)

Type:	Armoured scout car
Manufacturer:	DAF
Fighting weight:	5800 kg
Length:	4750 mm
Breadth:	2080 mm
Height:	2160 mm
Motor:	Ford mercury 8-cylinder petrol engine
Cubic capacity:	3918 ccm
Performance kW/hp:	71/96
Performance weight:	16.6 hp/t
Fuel capacity:	100 litres
Range:	320 km (road)
Crew:	5
Armament:	1 x 3.7 cm KwK, 3 x 7.92 mm MG
Armour: max.	12 mm
Fording depth:	0.5 m

At the end of 1934 the Dutch Army ordered 26 Type M36 and M38 armoured scout cars from the Swedish manufacturer Landsverk. These had Daimler-Benz and Büssing-NAG bogies and resembled the Sd.Kfz.231 (six wheeler). Most of these vehicles fell in to German hands in May 1940 and were later used for security purposes.

From 1939 another twelve armoured cars were acquired but this time from the local factory of the DAF firm. These M39's were of progressive design with an all-in-one body, thus dispensing with a conventional chassis. The five-man crew included a driver fore and rear. The Ford engine at the rear drove the two rear axles. Besides a 3.7-cm KwK and a 7.92 mm MG in the rotating turret, the cars had two other 7.92 mm MGs, one forward, the other at the rear. German troops captured eleven M39's and these served with Wehrmacht units into the USSR.

PzSpWg DAF 201(h) of 18.Infanterie Div., 1940. The DAF machine-gun barrels were given large size protective jackets. The roller ahead of the leading wheel was to prevent bogging down on slopes. (Vincent Bourguignon)

Panzerspähwagen BAF 203 (r)

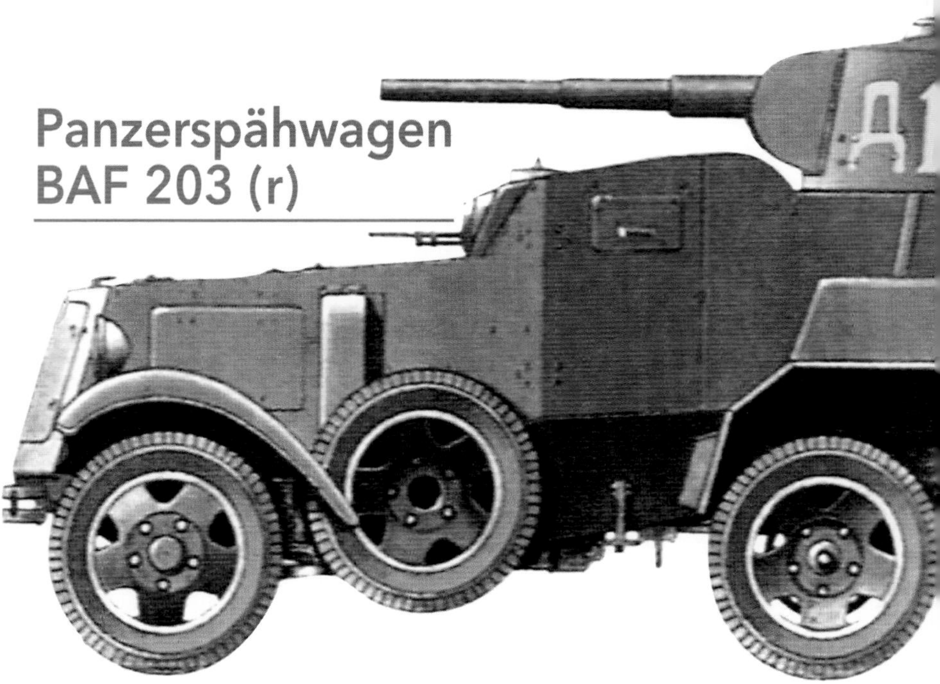

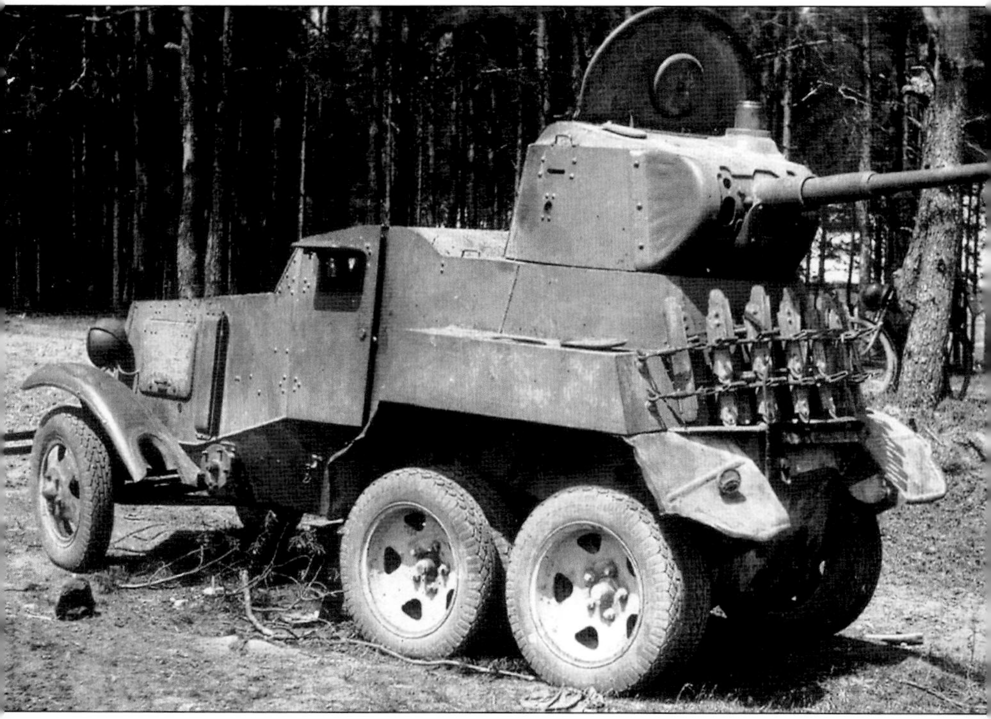

A Red Army BA-10M, Maikop area 1942 (WKA)

Type:	Armoured scout car (details for BA-10)
Manufacturer:	Soviet State
Fighting weight:	5140 kg
Length:	4650 mm
Breadth:	2070 mm
Height:	2210 mm
Motor:	GAZ M-1 4-cylinder petrol engine
Cubic capacity:	3300 ccm
Performance kW/hp:	37/50
Performance weight:	9.7 hp/t
Top speed:	55 km/hr (road)
Fuel capacity:	Not known
Range:	300 km (road)
Crew:	4
Armament:	1 x 4.5 cm KwK K20, 2 x 7.62 mm MG DT
Armour:	4 to 15 mm
Fording depth:	0.6m

In the early months following the attack on the Soviet Union German troops captured a large number of armoured scout cars, all of which were outwardly similar and based on a modified 6 x 4 Ford (GAZ-AAA) lorry bogie. Designated BA-3, BA-6 and BA-10 in the Red Army, in German service they were all tagged BAF 203(r). BA-10 was the last in a series which improved during the course of production. The BA-3 of 1934, and BA-6 of 1936 all had major drawbacks and apart from the efficient 4.5-cm gun were thought of little use even by the Red Army. This was reflected in the production output; thus 169 BA-3 were built, 386 BA-6 and 3311 BA-10. Common to all was the engine situated forward, an MG in a splinter shield to one side of the driver, free turning spare wheels to prevent bogging down in the terrain and chains for the rear wheels as an aid to traction. Nevertheless the cross-country performance of all models was inadequate for the lack of all-wheel drive, an underpowered 50 hp engine and heavy ground pressure.

The 1938 basic version of the BA-10 was armed with a 3.7 cm KwK replaced in 1939 by a 4.5 cm weapon, (BA-10a). Another but rare variant was the BA-10ZD able to run on railway metals. Production of the BA-10M ended in November 1941.

The Wehrmacht used numerous vehicles of this series, frequently they were employed by those same units which had captured them. Occasionally whole platoons of these armoured scout cars were fitted with captured materials. The BAF 203(r) was also used for security tasks and anti-partisan work in rearward areas.

Eastern Front, summer 1941. A recently captured BA-10 with lateral spare wheels missing. The chains for the rear wheels are secured at the rear.

Schwerer Panzerspähwagen Sd.Kfz. 231 (eight-wheeler)

From the outset, Army High Command knew of the limited capabilities of the six-wheel scout car. Based on good results from 1927 in the course of testing eight- and ten-wheeler vehicles, from 1934 to 1935 Büssing-NAG was commissioned by the Army Weapons Department to produce a car of this type. The basis was a "GS" bogie of 4120 kg, the eight wheels of which were suspended individually from longitudinal leaf-springs, and had all-wheel drive and controls. Travel forwards or backwards could be taken over by a driver facing in the chosen direction. The special gearing developed by Büssing made it possible to proceed in either direction at the same

speed. The 150-hp Büssing L 8 V motor was situated at the rear. The cubic capacity of the engine was initially 7913 ccm, but during the course of production the cylinder bore was enlarged raising the cubic capacity to 8363 ccm and the engine output to 180 hp.

The armoured superstructure of the new family of heavy reconnaissance vehicles - along with an armoured scout car the Army weapons department planned to also have a radio and command car - was developed at Kiel by Deutsche Werke. Its typically angular superstructure seen on the German armoured scout cars of the time was welded,

Opposite Top: Western Front, 1940. (WKA)

Opposite Bottom: This vehicle of 17.Panzer Division has the so-called "anti-tank protection" fitted at the front.

Sd.Kfz.231 (8-wheeler) of an unknown unit, Poland, September 1939. (WKA)

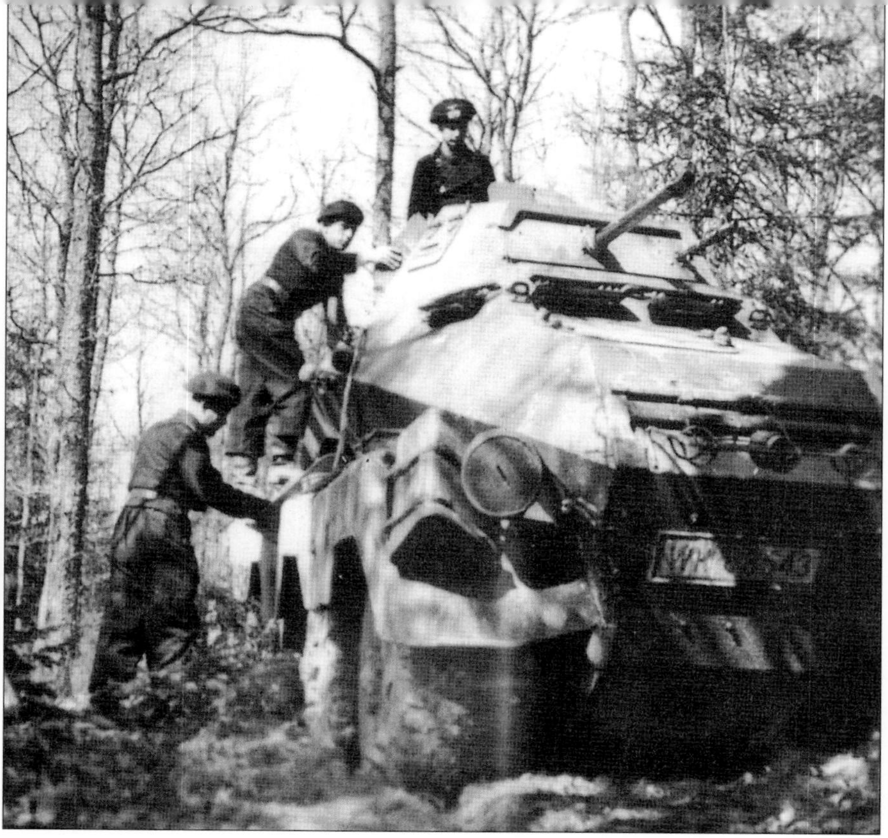

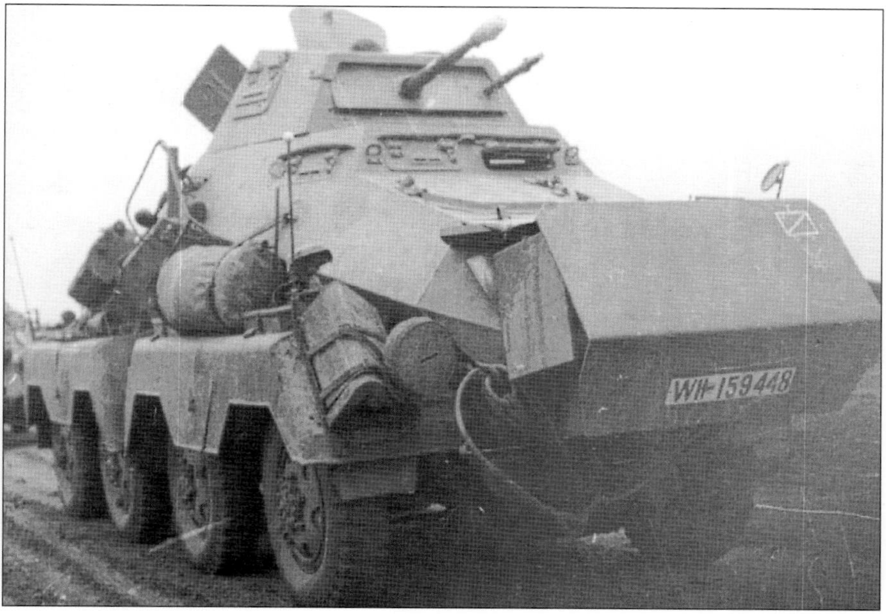
WH-159448

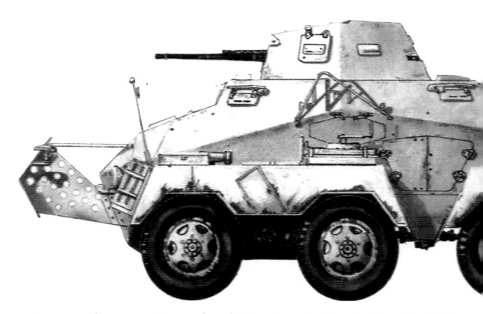

Heavy armoured scout car in winter camouflage of SS-Panzer Grenadier Division Das Reich, winter 1942/43 Easter Front. (WKA)

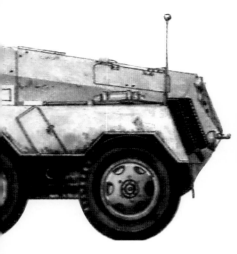

Type:	Heavy armoured scout car
Manufacturer:	Final assembly Schichau
Fighting weight:	8300 kg
Length:	5850 mm
Breadth:	2200 mm
Height:	2340 mm
Motor:	Büssing-NAG L 8 V, V-8 pettrol engine
Cubic capacity:	7913 ccm
Performance kW/hp:	110/115
Performance weight:	18 hp/t
Top speed:	90 km/hr (road)
Fuel capacity:	150 litres
Range:	300 km (road 160 km (terrain)
Crew:	4
Armament:	1 x 2 cm KwK 30/38 L/55, 1 x 7.92 mm MG 34
Armour:	5 to 14.5 mm (later up to 30 mm)
Fording depth:	1 m

14.5 mm thick at the front and 8 mm at the sides and rear. The floor and ceiling were protected by 5 mm thick plating. Access into the vehicle was by two-part hatches in front of the driver and laterally through the superstructure.

Armament was a 2-cm KwK 30 (later 38) for which 180 rounds were available and a co-axial 7.92 mm MG 34 with 2100 rounds in a rotating turret. The hand-operated turret with hatches in the roof and rear was occupied by the commander and gunner. After a protoype Vs Kfz.623 had been thoroughly tested successfully, series production began at Schichau/Elbing, and the first examples were delivered to the Wehrmacht in 1937. Because the vehicles were assuming the same role as the former 6-wheel armoured cars, they were given the same Sd.Kfz. designation but with the suffix "(8-wheel)". Further changes in detail followed in the course of production. Together with an increase in armour and

motor performance the hatches, vision flaps and the covers for the cooling grills were modified.

Operationally the vehicles proved that the GS-bogie gave them outstanding cross-country abilities. The bogie was complex, however and visits to the repair yard were common. Nevertheless this type of scout car belonged to the best of its class. As the war went on it was seen that the GS series was very vulnerable by reason of its large size and poor armour, and therefore from 1940 the vehicles received initially a so-called "anti-tank projectile protection" (Pak-Schutz), a 10 mm thick armour plate placed some distance forward of the vehicle front. Later armour protection was increased to 30 mm. Up to 1941 the vehicles of this type rendered sterling service, but North Africa and the USSR showed their limitations, and for this reason production was halted at the end of 1942. Nevertheless the type remained operational until 1945. A total of 1,235 vehicles of the eight-wheel armoured scout car series (Sd.Kfz.231/232/233/263) were built.

This artillery measuring vehicle on the basis of the Sd.Kfz.231(8-wheeler) captured by the US Army at the Hillersleben firing range in 1945 was probably a one-off. (US Army).

Schwerer Panzerspähwagen (Fu) Sd.Kfz. 232 (eight-wheeler)

Although the basic version of the Sd.Kfz.231 (8-wheeler) was fitted with radio, in this construction series there was another special radio car with a more powerful radio installation the purpose of which was to remain in touch with the command levels during long-range reconnaissance missions. Outwardly both models were to be distinguished from the SdKfz.231 by the huge frame aerial, the bogie and armoured superstructure being identical with the basic version. The frame aerials made a good target and during the war were often replaced by telescopic or radial aerials.

Crew numbers and armament of both versions were the same but the Sd.Kfz.232 only carried 1,500 rounds for the co-axial MG 34 in the turret. From 1940 the Sd.Kfz.232 was given a crash-plate ("anti-tank projectile protection") ahead of the lower face section. In the course of production the same

Sd.Kfz.232 (8-wheeler), Aufkl.Abt.33 (reconnaissance detachment), Marsa Matruh (Egypt) 1942. (WKA)

The access hatches of the "GS Reihe" can be seen between the second and third axles in this pre-war photo.

changes were made to the 232 as to the 231. Series production of the Sd.Kfz.232 halted at Deutsche Werke in May 1940 but was resumed in June 1941. Manufacture of the Sd.Kfz.232 (8-wheeler) was finally terminated at the end of 1942.

Type:	Heavy armoured radio scout car (details for late version)
Manufacturer:	Final assembly Deutsche Werke
Fighting weight:	8800 kg
Length:	5850 mm
Breadth:	2200 mm
Height:	2900 mm (with aerial)
Motor:	Büssing-NAG L 8 V-GS, V-8 petrol engine
Cubic capacity:	8360 ccm
Performance	kW/hp: 132.5/180
Performance weight:	20.5 hp/t
Top speed:	90 km/hr (road)
Fuel capacity:	150 litres
Range:	300 km (road) 160 km (terrain)
Crew:	4
Armament:	1 x 2 cm KwK 30/38 L/55, 1 x 7.92 mm MG 34
Armour:	5 to 30 mm
Fording depth:	1 m

Heavy armoured radio cars of the Panzer Grenadier Division Grossdeutschland on low loaders heading for the front, Russia 1942. (WKA).

Schwerer Panzerfunkwagen Sd.Kfz. 263 (eight-wheeler)

Type:	Heavy armoured radio car
Manufacturer:	Büssing-NAG, Deutsche Werke
Fighting weight:	8800 kg
Length:	5850 mm
Breadth:	2200 mm
Height:	2900 mm
Motor:	Büssing-NAG L 8 V, V-8 petrol engine
Cubic capacity:	7913 ccm
Performance kW/hp:	110/150
Performance weight:	17 hp/t
Top speed:	90 km/hr (road)
Fuel capacity:	150 litres
Range:	300 km (road) 160 km (terrain)
Crew:	4
Armament	1 x 7.92 mm MG 34
Armour:	5 to 18 mm
Fording depth:	1 m

In common with the six-wheeler versions, the eight-wheelers had an armoured radio car version, the Sd.Kfz.263 (8-wheeler). These saw service mainly in the signals sections of the panzer and motorized divisions to maintain contact with their own reconnaissance vehicles. Now and again divisional commanders used them as command vehicles for their comprehensive radio installation. In this respect Rommel may be mentioned, who during the campaign in the West directed 7.Panzer Division for some time from a heavy armoured radio car of this type.

In place of a rotating turret the Sd.Kfz.263 had a fixed superstructure enabling the fitting of map tables and an efficient 100-Watt FuG-11 radio installation with a range of up to 200 kilometres. Besides a frame aerial, behind the crew space was a telescopic aerial which could be hand-cranked up to a height of 9 metres. During the course of the war these were replaced in part by so-called umbrella aerials. The only armament was an MG 34 in a

The armoured radio car on operations. (WKA)

splinter shield in the upper face of the superstructure for which 1050 rounds of ammunition was carried. Production of this costly special vehicle was halted in 1941, the role being taken over by half-tracks with corresponding equipment. Nevertheless some of these cars were still at the front in 1944.

On this heavy armoured radio car the telescopic aerial can be seen at the rear. The white cross on the superstructure indicates that the vehicle took part in the Polish campaign.

Schwerer Panzerspähwagen Sd.Kfz. 233 (eight-wheeler)

Type:	Heavy armoured scout car
Manufacturer:	Büssing-NAG
Fighting weight:	8580 kg
Length:	5850 mm
Breadth:	2200 mm
Height:	2250 mm
Motor:	Büssing-NAG L 8 V, GS V-8-petrol engine
Cubic capacity:	8360 ccm
Performance kW/hp:	132.5/180
Performance weight:	21 hp/t
Top speed:	90 km/hr (road)
Fuel capacity:	150 litres
Range:	300 km (road) 160 km (terrain)
Crew:	3
Armament:	1 x 7.5 cm StuK 37 L/24, 1 x 7.92 mm MG 42
Armour:	5 to 30 mm
Fording depth:	1 m

The Sd.Kfz.233 was basically a turret-less variant of the Sd.Kfz.231. Replacing the 2-cm KwK this version had a short-barrelled 7.5-cm StuK 37 L/24 as also appeared in the earlier versions of the PzKpfw IV and StuG III. The Stuk (Sturmkanone) was mounted well forward in the open top armoured superstructure and had a very limited traverse of 12° to either side. The gun was aimed by a a simple pointer-sight. The vehicle carried 32 rounds which included hollow charge anti-tank ammunition. A 7.92 mm MG 42 with 1,500 rounds was carried in the crew room. Because the weapons and ammunition took up so much space, the crew was limited to three, namely driver, rear-facing driver/radio operator and gunner/commander. The Sd.Kfz.233 built at Büssing-NAG at Leipzig in 1942 were mainly conversions from the Sd.Kfz.231 and were used principally as mobile fire support for scout cars attached to armoured reconnaissance units.

Sd.Kfz. 233. (US Army)

Polizei-Panzer-kampfwagen ADGZ

Type:	Police armoured car
Manufacturer:	Steyr
Fighting weight:	12 tonnes
Length:	6260 mm
Breadth:	6260 mm
Height:	2560 mm
Motor:	Austro-Daimler M612 6-cylinder petrol engine
Cubic capacity:	11,946 ccm
Performance kW/hp:	110/150
Performance weight:	12.5 hp/t
Top speed:	70 km/hr (road)
Fuel capacity:	200 litres
Range:	400 km (road) 225 km (terrain)
Crew:	6
Armament:	1 x 2-cm KwK 35 L/45, 2 or 3 x 7.92 mm MG 34
Armour: max.	11 mm
Fording depth:	1 m

The development of this armoured car began at the Austro-Daimler works from 1931 and was taken over in 1934 by Steyr. By 1937, twenty-seven of these vehicles known as M-35 had been delivered to the Austrian army and police. After the annexation of Austria in March 1938 these cars went to German SS and police units. The police armoured car ADGZ had driver seats for both directions and thus had no front or rear in the true sense, it could be driven at the same speed going either way. Armament was a 20-mm KwK and one MG in the turret plus one or two in the bodywork. During the war vehicles of this type were used primarily for police duties and anti-partisan work. In 1942 Steyr delivered another 25 to the Waffen-SS who used them mainly in the Balkans and Russia for this purpose. As with other scout cars, the ADGZ also had a version with frame aerial.

An armoured car of the ADGZ type were used principally by SS and police units for security purposes. (WKA)

Schwerer Panzerspähwagen Sd.Kfz. 234/1 (2 cm)

Type:	heavy armoured scout car
Manufacturer:	Büssing-NAG
Fighting weight:	11,500 kg
Length:	6 metres
Breadth:	2330 mm
Height:	2100 mm
Motor:	Tatra 103 V-12 diesel engine
Cubic capacity:	14,825 ccm
Performance	kW/hp: 162/220
Performance weight:	19.1 hp/t
Top speed:	90 km/hr (road)
Fuel capacity:	360 litres
Range:	900 km (road), 600 km (terrain)
Crew:	4
Armament:	1 x 2 cm KwK 38,
	1 x 7.92 mm MG 42
Armour:	5 to 30 mm
Fording depth:	1.2 m

While production of the GS-car was still continuing, in 1940 Büssing-NAG received from the Army Weapons Department the contract to develop its successor. In order to keep the height of the vehicle as low as possible, Büssing designed a new all-in-one body, thus abandoning the conventional framework chassis such as the GS-bogie. Front armour was increased to 30 mm, at the sides and rear it remained 8 mm and 10 mm respectively. This new model known as "ARK" had the proven characteristics of its predecessor, for example independent suspension, all-wheel drive and control, and the ability to proceed forwards or backwards at the same speed. Larger and broader tyres increased substantially the performance cross-country. Emphasis at the development stage was laid on long range, therefore an air-cooled diesel was installed to put an end to the cooling problems arising from the GS-series having the engine at the rear.

The firm Tatra handled the engine. The first ARK prototypes were tested in mid-1941 and displayed to the Army Weapons Department at the end of the year. Difficulties with the engine, particularly heavy vibrations and noise, and the desire of the Army Weapons Department to want it to withstand tropical conditions, resulted in the series run not being taken up for around two more years.

The rotatable turret of the Sd.Kfz.234/1 basic model was developed by Daimler-Benz and Schichau and resembled that of the Sd.Kfz.222 being open-topped but closed if required by folding grilles. A 2-cm KwK 38 (250 rounds) and a co-axial MG 42 (2400 rounds)

were so positioned as to be capable of elevation to +75° as flak. For transmission of reconnaissance reports the Sd.Kfz.234/1 had an FuG 12 SE 80 and a FuSprGer "f" radiophone with rod aerial. By the end of 1944 Büssing-NAG had turned out two hundred 234/1. Plans to equip the type with a 2-cm flak in a suspended chassis or with a triple-barrelled MG (as with the Sd.Kfz.251/21) were never realized.

(p.62): An Sd.Kfz 234/1 captured by the British. A part of the 2 cm KwK barrel seems to be missing. (Tank Museum)

Photo (p.63) Overhead view of the same scout car showing its hexagonal open turret. (Tank Museum)

Schwerer Panzerspähwagen Sd.Kfz. 234/2 (5 cm) Puma

The Sd.Kfz.234/2 Puma was probably the best known of all the armoured scout cars of the Wehrmac ht. Bodywork and bogie were identical with the Sd.Kfz.234/1 but the rotatable turret and armament were distinct. The closed upper rotatable turret was designed originally for the light panzer *Leopard* discontinued at the prototype stage. The Puma armament was the 5-cm KwK 39/1 L/60 with 55 rounds also used for the Pz.Kpfw.III version J. This gun had been developed from the Pak 38 and with the panzer shell 39 had a muzzle velocity of 823 m/sec. This would go through 78 mm armour at 500 metres. The KwK 39/1 thus gave the Puma a good chance against enemy light tanks and reconnaissance vehicles. A co-axial 7.92 mm MG 42 was fitted alongside the main weapon in the massively armoured "pig's head" and pot shield 100 mm thick. 2850 rounds were carried for the MG. Either side of the turret thrower cups were installed for smoke- or explosive projectiles. The crew was four men: driver, commander, gunner and rear-driver/radio operator. Radio was an FuG 12 SE 80 and radio phone FuSprGer "a". The first of these armoured cars were delivered in September 1943, by 1945 one hundred and one had been built. In concept and design the Sd.Kfz.234/2 was a landmark and influenced armoured car development long after 1945.

The Sd.Kfz. 234/2 was the only Wehrmacht armoured scout car type to receive a suggestive name. (Tank Museum)

(centre) A convoy of apparently brand new Sd.Kfz. 234/2 in 1944. (WKA)

(bottom) In all only 101 Pumas were built. (Tank Museum)

Type:	armoured scout car
Manufacturer:	Büssing-NAG
Fighting weight:	11,700 kg
Length:	6700 mm (with barrel)
Breadth:	2330 mm
Height:	2380 mm
Motor:	Tatra 103 V-12 diesel engine
Cubic capacity:	14,825 ccm
Performance kW/hp:	162/220
Performance weight:	18.8 hp/t
Top speed:	90 km/hr (road)
Fuel capacity:	360 litres
Range:	900 km (road), 600 km (terrain)
Crew:	4
Armament:	1 x 5 cm KwK 39/1 L760, 1 x 7.92 mm MG 42
Armour:	5 to 30 mm (turret screen 100 mm)
Foring depth:	1.2m

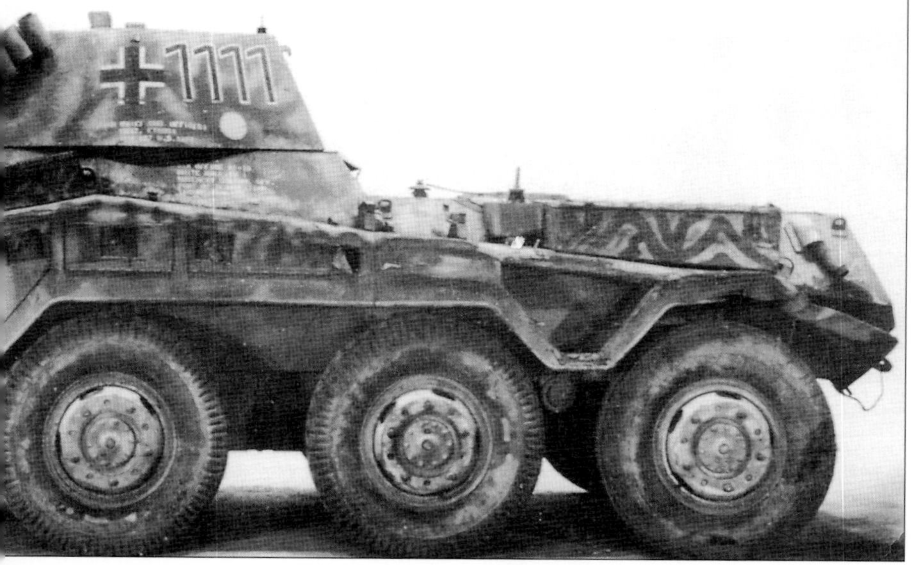
1111

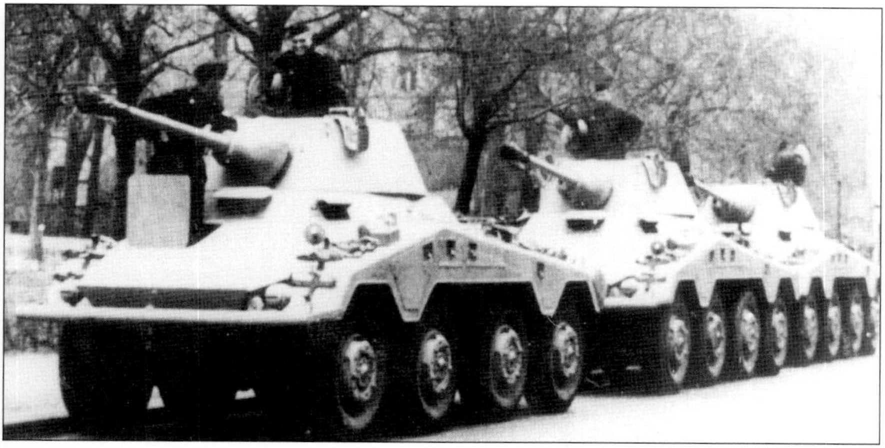

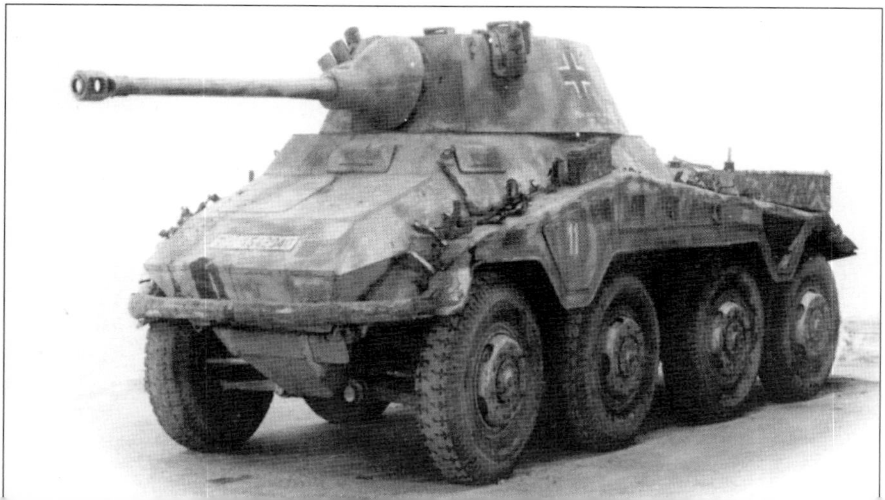

Schwerer Panzerspähwagen Sd.Kfz. 234/3 (7.5 cm)

The third version of the ARK-series was the Sd.Kfz.234/3. It was also developed with the idea of providing fire support for reconnaissance units equipped with the Sd.Kfz.234/1, for which purpose all armoured reconnaissance detachments were to have a platoon with six cars of this type. The bogie was identical with that of the Sd.Kfz.234/1, a 7.5 cm

All three photographs of the Sd.Kfz. 234/3 are of a scout car captured by US forces and investigated at the US Army Armour School, Fort Knox. (US Army)

KwK 51 L/24 with limited traverse of 12° to either side being fitted into the open armoured superstructure together with a 7.92 mm MG 42. 50 rounds were carried for the gun, 1,950 for the MG. Although the KwK 51 fired hollow charge munition as well as AP, smoke and shrapnel shells, it was not successful against enemy tanks by reason of the low muzzle velocity and the resulting strongly curved trajectory. The side walls of the superstructure were given height for crew protection. The radio installation was an FuSprGe "f" radiophone. Instead of the on-board radio-com as on the Sd.Kfz.234/1 and /2, this version only had a speech tube. The Sd.Kfz. 234/3 resembled the Sd.Kfz.233 very closely, but as with all ARK-car versions the wheel guards were continuous and not divided up as for earlier GS-series cars. Large stowage boxes were fitted into the wheel guards for tools, kit and crew effects. Only eighty-eight Sd.Kfz.234/3 were manufactured during the war.

Type:	Heavy armoured scout car
Manufacturer:	Büssing-NAG, Deutsche Werke
Fighting weight:	11,500 kg
Length:	6 metres
Breadth:	2330 mm
Height:	2210 mm
Motor:	Tatra 103 V-12 diesel engine
Cubic capacity:	14,825 ccm
Performance kW/hp:	162/220
Performance weight:	19.1 hp/t
Top speed:	90 km/hr (road)
Fuel capacity:	360 litres
Range:	900 km/hr (road) 600 km (terrain)
Crew:	4
Armament:	1 x 7.5 cm KwK 51 L/24, 1 x 7.92 mm MG 42
Armour:	5 to 30 mm
Fording depth:	1.2 m

Schwerer Panzerspähwagen Sd.Kfz.234/4 (7.5 cm Pak 40)

The fourth Sd.Kfz. 234 variant, also known as the "Pak-waggon", was made for use primarily in the anti-tank role and built on Hitler's personal initiative. He had suggested re-locating the powerful 7.5-cm Pak 40 L/46 in the crew room of the Sd.Kfz.234. Towards the end of the war this weapon was the standard Pak of the Wehrmacht anti-tank units and was also used on other self-propelled chassis, e.g. the tank hunters of the Marder series. The 6.8 kg panzer shell 39 had a muzzle velocity of 792 m/sec, and fired by the Pak 40 it would penetrate armour 89 mm thick at a range of 1000 metres. Development of the first prototypes concluded at the end of November 1944. The Pak 40 together with upper mount and splinter shield was located in the open-top crew area and had a limited traverse. Only 12 rounds could be carried. An MG 42 (1950 rounds) was also fitted in the crew area. Although basically only a temporary solution (and the gun weighed down heavily on the vehicle), the Sd.Kfz.234 proved a strong fighting machine and very mobile. Hitler was convinced that he had personally created one of the best anti-tank vehicles of the war and wanted a monthly production of 100. Only eighty-nine had been turned out by the capitulation, however. How many of these reached the front is not clear. The production figures for the Sd.Kfz.234 range vary widely. Büssing-NAG stated that by the war's end they had completed around 2,300 armoured scout cars of the ARK-series.

Sd.Kfz. 234/4 in the Panzermuseum, Munster. (Sebastian Hoppe)

Type:	Heavy armoured scout car
Manufacturer:	Büssing-NAG, Deutsche Werke
Fighting weight:	11,500 kg
Length:	6 metres (without barrel)
Breadth:	2330 mm
Height:	2210 mm
Motor:	Tatra 103 V-12 diesel engine
Cubic capacity:	14,825 ccm
Performance kW/hp:	162/220
Performance weight:	19.1 hp/t
Top speed:	90 km/hr (road)
Fuel capacity:	360 litres
Range:	900 km (road) 600 km (terrain)
Crew:	4
Armament:	1 x 7.5 cm Pak 40 L/46, 1 x 7.92 mm MG 42
Armour:	5 to 30 mm
Fording depth	1.2 m

Side profile illustration of the Sd.Kfz 234/4.
(Vincent Bourguignon)

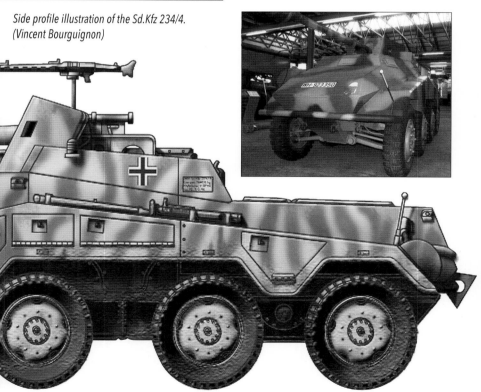

Leichter Schützenpanzer-wagen Sd.Kfz. 250/1

During the development of the medium armoured personnel carrier Sd.Kfz.251 designed to carry a whole platoon of grenadiers, the Army Weapons Department decided to have a car for a half-platoon. The basis was to be the DEMAG (Wetter/Ruhr) 1-tonne tug (Sd.Kfz.10). Therefore DEMAG received the contract to redesign its "D 7" bogie for the purpose. The main changes were a shortening of the bogie and reduction by two axles, and a reduction of the cooling system, fuel tanks, the steering assembly and the exhaust. Büssing-NAG designed the armoured superstructure of the new light armoured personnel carrier (*German* SPW, English APC). The front armour was 14.5 mm, sides and rear 8 mm, floor and ceiling 5 mm. Basically the technical concept and structure was that for the Sd.Kfz.251. The motor was forward, behind it lay the open-top crew space with a single flap rear door. The crew space could be covered over by tarpaulin, three hoops being fitted for this purpose. The ten running wheels with rubber tread were torsion sprung. Drive was on the leading axle of the tracked wheels, not on the tyred front axle used only for steering. For major changes of direction a track-brake could be employed. In contrast to the Sd.Kfz.251 the light SPW/APC had a semi-automatic Maybach gear system.

After several prototypes had been completed and tested in 1939, the first series vehicles went into action at the front during the French campaign in 1940. Construction of both light and medium APC's went ahead during the war at a number of firms because of the great demand for this class of vehicle. DEMAG and Mechanische Werke Cottbus completed the bogies while the armoured superstructure was

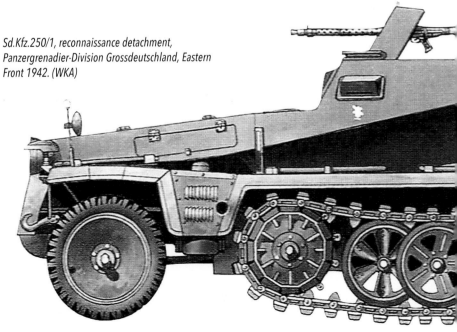

Sd.Kfz.250/1, reconnaissance detachment, Panzergrenadier-Division Grossdeutschland, Eastern Front 1942. (WKA)

Two Sd.Kfz.250/1 of Panzergrenadier-Division Grossdeutschland, Eastern Front 1942. (WKA)

undertaken by Deutscher Edelstahl AG, Steinmüller, Böhler and a series of other firms. Evens & Pistor, Weserhütte, Wumag, Wegmann and others handled the final assembly.

There were two version of the Sd.Kfz.250/1. One carried four soldiers and was armed with two light MG 34 or MG 42 and 2010 rounds of ammunition. One of these was given a splinter shield, the other was located in the crew space or mounted on a flak arm. The second version had a single MG on a tripod fixed outside the car.

Type:	Light armoured personnel carrier
Manufacturer:	Final assembly at Evens & Pistor, Weserhütte, Wumag; Wegmann and others
Fighting weight:	5700 kg
Length:	4560 mm
Breadth:	1950 mm
Height:	1660 mm
Motor:	Maybach HL 42 TRKM, 6-cylinder petrol engine
Cubic capacity:	4170 ccm
Performance kW/hp:	73.6/100
Performance weight:	17.5 hp/t
Top speed:	65 km/hr (road)
Fuel capacity:	140 litres
Range:	350 km (road) 200 km (terrain)
Crew:	2 + 4
Armament:	2 x 7.92 mm MG 34 or MG 42
Armour:	6 to 14.5 mm
Fording depth:	0.7 m

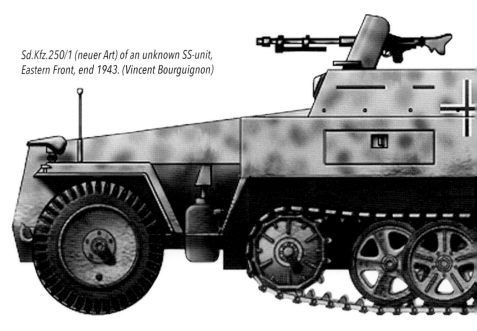

Sd.Kfz.250/1 (neuer Art) of an unknown SS-unit, Eastern Front, end 1943. (Vincent Bourguignon)

Leichter Schützenpanzerwagen (neuer Art) Sd.Kfz. 250/1

Up to the autumn of 1943 only minor modifications were made to the Sd.Kfz.250 during production in order not to delay the assembly line. From then on the superstructure of the light APC was greatly simplified. The number of armoured plates required in manufacture was halved, and these were now flat instead of angular, the thickness remaining the same except for the one-piece front plating now 15 mm instead of the previous 14.5 mm. Stowage lockers were integral at the sides of the superstructure to allow the crew more space for belongings. The armament was unchanged. The various versions of the Sd.Kfz.250 were not given numerals to distinguish the different modified types although occasionally the suffix "neuer Art" of "n/A" (=new type) was added. Sometimes the troops added a

5-cm Pak 38 in the crew space of the light APC n/A. An example like this can be seen today at the Military Museum, Belgrade.

Numerous variants of both versions of the Sd.Kfz.250 were produced. There are fourteen different versions of the light APC, diverse sub-variants, prototypes and unofficial deviations, some of which had parts of the Sd.Kfz.251 added. In addition to these variants came armoured observation cars, which also served to transport the Staffs of SP-gun units (Sd.Kfz.250/4), and artillery spotters (Sd.Kfz.250/12). By when production terminated at the end of 1944, around 6,800 Sd.Kfz.250 of all versions had been built, 3,590 of these in 1943 and 1944.

Saint Aubin d'Appenai (Normandy), summer 1944: US troops inspecting the wreckage of a light armoured personnel carrier "neuer Art" of Panzer-Aufklärungs-Abtg.2/2.Panzer-Division. (NARA)

Type:	Light armoured personnel carrier
Manufacturer:	Final assembly at Evens & Pistor, Weserhütte, Wumag; Wegmann and others
Fighting weight:	5380 kg
Length:	4610 mm
Breadth:	1950 mm
Height:	1660 mm (without MG-shield)
Motor:	Maybach HL 42 TURKM, 6-cylinder petrol engine
Cubic capacity:	4170 ccm
Performance kW/hp:	73.6/100
Performance weight:	18.6 hp/t
Top speed:	65 km/hr (road)
Fuel capacity:	140 litres
Range:	350 km (road), 200 km (terrain)
Crew:	2 + 4
Armament:	2 x 7.92 mm MG 34 or MG 42
Armour:	6 to 15 mm
Fording depth:	0.7 m

Leichter Fernsprechpanzer-wagen Sd.Kfz. 250/2

Few examples of this light APC variant were made and they differed outwardly from the basic version only in the cable drums mounted on the front mudguards. The vehicles were used to transport telephone-cable-layer troops who required armoured protection while at work. Besides the two drums on the mudguards a third drum was located in the crew space so that work could proceed laying three lines at the same time. This third drum could also be used to winch in cable. Crew was four, for self defence the *leichter Fernsprechpanzerwagen* (occasionally called *leichter Fernsprechbauwagen*, the difference in terminology having no significance) had a 7.92 mm MG 34 or 42 behind a splinter shield. Besides the three drums, cars of this kind carried comprehensive telephone equipment in the crew room and the outer stowage lockers.

Type:	Light telephone cable-layers armoured car
Manufacturer:	Final assembly at Evens & Pistor, Weserhütte, Wumag; Wegmann and others
Fighting weight:	5450 kg
Length:	4560 mm
Breadth:	1950 mm
Height:	1660 mm (without aerial)
Motor:	Maybach HL 42 TRKM, 6-cylinder petrol engine
Cubic capacity:	4170 ccm
Performance kW/hp:	73.6/100
Performance weight:	18.3 hp/t
Top speed:	65 km/hr (road)
Fuel capacity:	140 litres
Range:	350 km (road) 200 km (terrain)
Crew:	4
Armament:	1 x 7.92 mm MG 34 or MG 42
Armour:	6 to 14.5 mm
Fording depth:	0.7 m

Sd.250/2, 1.Panzer-Division, France May/June 1940. (Vincent Bourguignon)

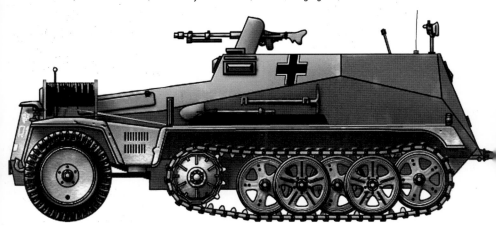

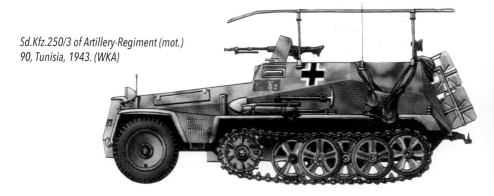

Sd.Kfz.250/3 of Artillery-Regiment (mot.) 90, Tunisia, 1943. (WKA)

Leichter Funkpanzerwagen Sd.Kfz.250/3

The light armoured radio car based on the Sd.Kfz.250 had a comprehensive radio installation and came in five sub-variants, differing in their combination of radio- and other signals equipment. Most of them carried initially the usual frame aerial, often replaced later by a rod or umbrella-aerial. Sd.Kfz.250/3 was also used by Luftwaffe units, e.g. as a radio car for flight liaison officers coordinating air and ground units. Together with general signals work the light armoured radio car was employed to maintain contact with the regiment or division and as a command car. The best-known Sd.Kfz.250/3 was assuredly Rommel's personal command car *Greif*, which he used in 1942 with the Africa Korps.

Sd.Kfz.250/3 of Panzer-Grenadier-Division Grossdeutschland, Eastern Front, 1942. (WKA)

Type:	Light armoured radio car
Manufacturer:	Final assembly at Evens & Pistor, Weserhütte, Wumag; Wegmann and others
Fighting weight:	5750 kg
Length:	4560 mm
Breadth:	1950 mm
Height:	1660 mm (without aerial)
Motor:	Maybach HL 42 TRKM, 6-cylinder petrol engine
Cubic capacity:	4170 ccm
Performance kW/hp:	73.6/100
Performance weight:	17.4 hp/t
Top speed:	65 km/hr (road)
Fuel capacity:	140 litres
Range:	350 km (road), 200 km (terrain)
Crew:	4
Armament:	1 x 7.92 mm MG 34 or MG 42
Armour:	6 to 14.5 mm
Fording depth:	0.7 m

Sd.Kfz.250/5 as artillery spotter, 21.Panzer-Division, North Africa, 1942. (WKA)

Sd.Kfz. 250/5-II was used mainly as a light armoured scout car fitted with an FuG 12 and FuSprGer "f" radiophone. The small size of this "light armoured reconnaissance car" made it ideal for the role. It was easy to hide and was rarely spotted due to its low height.

Leichter Beobachtungs-panzerwagen Sd.Kfz.250/5

Type:	Light armoured observation vehicle
Manufacturer:	Final assembly at Evens & Pistor, Weserhütte, Wumag; Wegmann and others
Fighting weight:	5700 kg
Length:	4560 mm
Breadth:	1950 mm
Height:	1660 mm (without aerial)
Performance kW/hp:	73.6/100
Performance weight:	17.5 hp/t
Top speed:	65 km/hr (road)
Fuel capacity:	140 litres
Range:	350 km (road) 200 km (terrain)
Crew:	4
Armament:	2 x 7.92 mm MG 34 or MG 42
Armour:	6 to 14.5 mm
Fording depth:	0.7 m

There were two versions of this model. Sd.Kfz.250/5-I was used primarily as a spotter vehicle with artillery and SP-gun units. Outwardly the cars were very similar to the basic model but for their observer role had necessary equipment such as the FuG 4 and FuG 8 SE 30 and a scissors optic. Earlier models had a very noticeable frame aerial, later generally exchanged for a 2-m high umbrella aerial. The

Sd.Kfz.250/5, Panzer-Artillerie-Regt.74/2.Panzer-Division, Eastern Front, 1943. (WKA)

Leichter Munitionspanzerwagen Sd.Kfz. 250/6

Type:	Light armoured ammunition car (version A)
Manufacturer:	Final assembly at Evens & Pistor, Weserhütte, Wumag; Wegmann and others
Fighting weight:	5940 kg
Length:	4560 mm
Breadth:	1950 mm
Height:	1660 mm (without MG)
Motor:	Maybach HL 42 TRKM, 6-cylinder petrol engine
Cubic capacity:	4170 ccm
Performance kW/hp:	73.6/100
Performance weight:	16.8 hp/t
Top speed:	65 km/hr (road)
Fuel capacity:	140 litres
Range:	350 km (road), 200 km (terrain)
Crew:	2
Armament:	1 x 7.92 mm MG 34 or MG 42
Armour:	6 to 14.5 mm
Fording depth:	0.7 m

The light ammunition armoured car became the successor in the autumn of 1941 to the Sd.Kfz.252 munitions supply vehicle for mobile assault guns. In contrast to its predecessor the Sd.Kfz.250/6 had no specially developed armoured superstructure and was practically indistinguishable from the basic version of the light APC. In the open-top crew space however racks were installed for the transport of ammunition.

The Sd.Kfz.250/6 came in two versions. Version A carried up to 70 shells for the 7.5-cm StuK 37 L/24; version B 60 rounds for the more powerful 7.5-cm StuK 40 L/48. Version A weighed 5940 kg fully laden, version B 6090 kg. In action both versions often hauled the Special-trailer 32/A with up to thirty-six 7.5-cm shells. The two-man light armoured ammunition car was equipped with a 7.92 mm MG 34 or MG 42 with splinter shield and carried 2010 MG rounds.

Sd.Kfz.250/6. (WKA)

Leichter Schützenpanzerwagen (sGrW) Sd.Kfz. 250/7

This version of the light APC also came in two versions. The first carried a heavy 8-cm calibre mortar 34, 12 mortar bombs, base plate and the crew. As with the Sd.Kfz. 251/2, the mortar could be fired from within the vehicle, but if possible it was operated with the weapon set up outside. Other armament was a 7.92 mm MG 34 or Mg 42.

The second version of the Sd.Kfz.251/7 came as a transport carrying sixty-six 8-cm mortar bombs and an optical rangefinder. It was often used by the commander of the heavy platoon of a panzergrenadier batallion. It was because the Sd.Kfz.251 was really too large for the role of a mortar car and medium APC's were urgently required for other assignments that the Sd.Kfz 250/7 often replaced the larger vehicle in the role.

Type:	Light armoured personnel carrier
Manufacturer:	Final assembly at Evens & Pistor, Weserhütte, Wumag; Wegmann and others
Fighting Weight:	6100 kg
Length:	4560 mm
Breadth:	1950 mm
Height:	1660 mm (without aerial)
Motor:	Maybach HL 42 TRKM, 6-cylinder petrol engine
Cubic capacity:	4170 ccm
Performance kW/hp:	73.6/100
Performance weight:	16.4 hp/t
Top speed:	85 km/hr (road)
Fuel capacity:	140 litres
Range:	350 km (road) 200 km (terrain)
Crew:	5
Armament:	1 x 8-cm sGrW 34, 1 x 7.92 mm MG 34 or MG 42
Armour:	8 to 14.5 mm
Fording depth:	0.7 m

Sd.Kfz.250/7 of the 4th (heavy) platoon of a light armoured reconnaissance detachment, Eastern Front 1942. (Vincent Bourguignon)

Leichter Schützenpanzer-wagen (7.5-cm) Sd.Kfz. 250/8

Type:	Light armoured personnel carrier
Manufacturer:	Final assembly at Evens & Pistor, Weserhütte, Wumag; Wegmann and others
Fighting weight:	6300 kg
Length:	4610 mm
Breadth:	1950 mm
Height:	2070 mm (with StuK)
Motor:	Maybach HL 42 TURKM,6-cylinder petrol engine
Cubic capacity:	4170 ccm
Performance kW/hp:	73.6/100
Performance weight:	15.9 hp/t
Top speed:	60 km/hr (road)
Fuel capacity;	140 litres
Range:	350 km (road) 200 km (terrain)
Crew:	3
Armament:	1 x 7.5 cm StuK 37 or 51 L/24, 1 x 7.92 mm MG 34 or MG 42
Armour:	6 to 15 mm
Fording depth:	0.7 m

In the spring of 1943 a small number of Sd.Kfz 250/8's (known unofficially as "gun-waggons") were turned out on the "old" bogie before the line was discontinued. In the autumn of 1943 the production of the "new" version of the light APC was reinstated. In the role of support vehicle Sd.Kfz.250/8 received a 7.5-cm StuK 37 L/24 (from the autumn of 1944 a slightly modified StuK 51). The short-barrelled gun was mounted above the driver's level behind a 14.5 mm thick armour shield and had only a limited range of traverse. The lateral sides of the superstructure 10 mm thick were extended upwards to better protect the crew. The vehicle carried 20 rounds of 7.5-cm ammunition inside the vehicle together with an MG 34 or 42. The type was usually detached to the 4th platoon of a light armoured reconnaissance company in order to provide its scout cars with covering fire.

Sd.Kfz.250/8 of an unknown unit, Eastern Front, end of 1944. (Vincent Bourguignon)

Leichter Schützenpanzer-wagen (2 cm) Sd.Kfz. 250/9

Very soon after the attack on the Soviet Union it was clear that the usual wheeled vehicles used as armoured scout cars were poorly effective in the prevailing circumstances on the Eastern Front. Half-track vehicles proved much more suitable in the difficult terrain. Generaloberst Guderian therefore requested a scout car based on the light APC. The result of this development was the Sd.Kfz.250/9, a pre-series being presented in March 1942 and successfully tried out on the Eastern Front in the summer of that year. Series production began in the spring of 1943. In contrast to all other versions of the

light APC the crew space of the Sd.Kfz.250/9 was protected by plate armour. The first vehicles had the Sd.Kfz.222 turret with pedestal mount 38 for a 2-cm KwK 38 and a co-axial MG 34. During the course of production the turret was replaced by a new model, for example that of the armoured scout car Sd.Kfz.234/1. This hexagonal turret had a suspended mount 38-giving the 2-cm KwK and co-axial MG 42 a range of elevation from -4° to +70°.

Both turrets were open-top but could be closed off by folding screens for protection against hand grenades. A tarpaulin could be pulled over the crew space in bad weather. 100 rounds were carried for the gun and 1100 rounds for the MG. Reports were transmitted by an FuG 12 with rod aerial. Vehicles of this type remained operational throughout the war.

Type:	Light armoured reconnaissance car
Manufacturer:	Final assembly at Evens & Pistor, Weserhütte, Wumag; Wegmann and others
Fighting weight:	5900 kg
Length:	4560 mm
Breadth:	1950 mm
Height:	2160 mm (with turret)
Motor:	Maybach HL 42 TRKM, 6-cylinder petrol engine
Cubic capacity:	4170 ccm
Performance kW/hp:	73.6/100
Performance weight:	16.9 hp/t
Top speed:	65 km/hr (road)
Fuel capacity:	140 litres
Crew:	3
Armament:	1 x 2-cm KwK 38, 1 x 7.92 mm MG 34 or MG 42
Armour:	6 to 14.5 mm
Fording depth:	0.7 m

This Sd.Kfz.250/9 was captured by Allied troops in Italy (here under inspection by a British Army officer) and apparently used against its former owners by the Americans (white stars on bodywork). (NARA)

Because of the difficult conditions in the terrain and on unmade roads of the Eastern Front, armoured scout cars were mostly replaced by the Sd.Kfz 250/9. (WKA)

Leichter Schützenpanzerwagen (3.7 cm Pak) Ss.Kfz. 250/10

Type:	Light armoured personnel carrier
Manufacturer:	Final assembly at Evens & Pistor, Weserhütte, Wumag; Wegmann and others
Fighting weight:	5670 kg
Length:	4560 mm
Breadth:	1950 mm
Height:	1970 mm (with 3-7-cm Pak 36)
Motor:	Maybach HL 42 TRKM, 6-cylinder petrol engine
Cubic capacity:	4170 ccm
Performance kW/hp:	73.6/100
Performance weight:	17.6 hp/t
Top speed:	65 km/hr (road)
Fuel capacity:	140 litres
Range:	350 km (road) 200 km (terrain)
Crew:	4
Armament:	1 x 3.7 cm Pak 36 L/45
Armour:	6 to 14.5 mm
Fording depth:	0.7 m

Analogous to the Sd.Kfz.251/10, this was a light armoured personnel carrier fitted with a 3.7-cm Pak 36 gun as a platoon leader vehicle giving fire support. As with the medium APC the Pak was also mounted less wheels and chassis tail on the armoured roof above the driver and radio operator. Traverse was 30° right and left and elevation -8° to +25° elevation. The splinter shield was often modified, the original proving too high. 216 rounds were carried for the gun, no MG was fitted. This version was one of the first light APC's and operational by 1940. Cars of this kind were only built on the basis of the "old" Sd.Kfz.250 version, and their production terminated at the end of 1943. Another version of the light APC armed with a gun was the Sd.Kfz.250/11: 168 rounds were carried for its heavy Panzerbüchse 41 (sPzB 41) with conical barrel (2.8/2 cm calibre). Also brought along was an MG 34 with 1100 rounds and at the outside rear of the vehicle a light field mounting for the sPzB, which could therefore be used independently. In North Africa, some Deutsches Afrika Korps vehicles replaced the German weapon by a French 2.5 cm Pak 112(f) anti-tank gun.

An Sd.Kfz.250/10 with damaged front axle captured by British troops in North Africa. (IWM)

Leichter gepanzerter Munitionstransportkraftwagen Sd.Kfz. 252

The Sd.Kfz.252, in common with the light APC, was based on the D 7 DEMAG bogie and was therefore another variant of the Sd.Kfz.250 family but with a closed-top armoured superstructure. The vehicle was also better armoured at the front (18 against 14.5 mm). The crew entered through two rear-opening single-flap hatches in the roof. The armour at the rear was sharply inclined. A large double door there gave access into the crew space. The purpose of these vehicles was to supply SP-assault guns with ammunition. In order to increase the amount conveyed, in action the vehicles frequently towed an unarmoured trailer (Sd.Anh.32/A), designed and manufactured by Wegmann, loaded with 36 rounds of 75 mm ammunition.

The Sd.Kfz.252 had a 7.92 mm MG 34 with 2010 rounds carried in the crew space. The development of the type began in 1937 but series production began only in 1940 at Wegmann of Kassel, the designer of the armoured superstructure. Only the final assembly was carried out at Wegmann. The bogie came from DEMAG, the armoured superstructure mainly from Böhler & Co.AG of Austria. The first vehicles reached the troops in June 1940. From January 1941

On operations the Sd.Kfz.252 very often pulled the ammunition trailer (Sd.Anh.32/A) with 36 shells. (WKA)

Deutsche Werke of Kiel also undertook final assembly work. In September 1941 the production of the Sd.Kfz.252 was halted again on the grounds that that kind of vehicle was too expensive bearing in mind its purpose. By then a total of 413 light armoured ammunition transporters had been built.

Type:	Light armoured munitions transporter
Manufacturer:	Final assembly at Wegmann, Deutsche Werke
Fighting weight:	5730 kg
Length:	4700 kg
Breadth:	1950 mm
Height:	1800 mm
Motor:	Maybach HL 42 TRKM, 6-cylinder petrol engine
Cubic capacity:	4170 ccm
Performance kW/hp:	73.6/100
Performance weight:	17.5 hp/t
Top speed:	65 km/hr (road)
Fuel capacity:	140 litres
Range:	350 km (road) 200 km (terrain)
Crew:	2
Armament:	1 x 7.92 mm MG 34
Armour:	8 to 18 mm
Fording depth:	0.7 m

Sd.Kfz.252 showing the inclined rear of the superstructure with double doors. Centre sector of the Eatsern Front, August/September 1943 (BA cc-by-sa 3.0).

Leichter gepanzerter Beobachtungs-kraftwagen Sd.Kfz. 253

Type:	Light armoured observation car
Manufacturer:	Final assembly Wegmann
Fighting weight:	5730 kg
Length:	4700 mm
Breadth:	1950 mm
Height:	1800 mm
Motor:	Maybach HL 42 TRKM, 6-cylinder petrol engine
Cubic capacity:	4170 ccm
Performance kW/hp:	73.6/100
Performance weight:	17.5 hp/t
Top speed:	65 km/hr (road)
Fuel capacity:	140 litres
Range:	350 km (road) 200 km (terrain)
Crew:	2
Armament:	1 x 7.92 mm MG 34
Armour:	8 to 18 mm
Fording depth:	0.7 m

The Sd.Kfz.253 was also based on the DEMAG D 7 bogie and at first glance looked like an Sd.Kfz.250. However, the Wegmann-designed armoured superstructure was closed on top. In the roof was a large two-flap observation hatch with another small aperture for raising the scissors optic. The four-man

Armoured observation vehicle during the Wegmann trials. (WKA)

Sd.Kfz.253 of an unknown StuG detachment, Stalingrad area, October 1942 (WKA)

crew entered by a second angular hatch at the rear of the superstructure and a single-flap rear door on the left side.

The Sd.Kfz.253 was thought of as a support vehicle for SP-assault gun units and to function primarily as an observation and command post. For this purpose it was equipped with FuG 15 and 16 radio installations (and a removable back pack radio), the 2-metre rod aerial being located at the rear of the superstructure right side. This could be folded down if necessary along a wooden protective duct projecting forward of the driver's position.

As with the Sd.Kfz.252 the concept and development of the light armoured observation car began in 1937, but series production did not ensue until March 1940. The first of these vehicles were operational during the campaign in France in the months of May and June 1940. Later the type served mainly on the Eastern Front. The same fate befell the Sd.Kfz.253 as did the 252 after it was soon realized that variants of the light APC (Sd.Kfz.250/4 and 250/5) could do the job of the Sd.Kfz.253 at a much more favourable price and so the production ended in June 1941. A total of only 285 Sd.Kfz.253 left the Wegmann assembly lines.

Sd.Kfz.253 cross-country, Bulgaria, April/May 1941. The rod aerial and its protective duct above the driver's visor can be clearly seen. (BA cc-by-sa 3.0)

Leichter Schützenpanzerwagen U304 (f)

Type:	Light APC (dimensions of version 1)
Manufacturer:	Unic, Baukommando Bauer
Fighting weight:	5400 kg
Length:	4850 mm
Breadth:	1800 mm
Height:	1950 mm (without MG)
Motor:	Unic P 39, 4-cylinder petrol engine
Cubic capacity:	3450 ccm
Performance kW/hp:	44/60
Performance weight:	11.1 hp/t
Top speed:	45 km/hr (road)
Fuel capacity:	160 litres
Range:	400 km (road) 160 km (terrain)
Crew:	2 + 8
Armament:	2 x 7.92 mm MG 34 or MG 42
Armour:	Max. 10 mm
Fording depth:	0.8 m

At the beginning of the campaign in the West in May 1940, French forces had over 3,276 light traction vehicles of the type Unic P107. These non-armoured half tracks designed originally by Citroën served in numerous roles, for example as artillery tugs, crew or ammunition transports and in pioneer units. The Unic-car tracks were much simpler than those of German traction vehicles or APCs. In place of a torsion-bar sprung bogie with smeared crawler tracks of steel and exchangeable rubber cushioning, the French half-tracks had a Citroën-Hinstin-Kégresse bogie. Its track was a single reinforced rubber belt, suspension being undertaken by longitudinal leaf springs. Even the bogies of US half-tracks were based on this system later. After the defeat of France in 1940, German forces captured a large number of Unic P107 tractor units, albeit seriously damaged. The vehicles were given a general overhaul by French industry and entered Wehrmacht service under the designation *leichter Zugkraftwagen* P107 U304(f) and used as originally intended by the French. In

1943 and 1944, however, a large number of them were converted into light APC's and given armoured superstructures by Baukommando Becker working in collaboration with the Army Weapons Department (Paris foreign branch) and elements of French industry under German instructions and supervision.

There were two versions. The first was very simple and had upright sides, the other was angular with inclined side walls similar to the Sd.Kfz.250 or 251.

*SPW (APC) U304(f).
(Vincent Bourguignon)*

Both versions had open tops, although on some of the second version the top was closed. Along with driver and co-driver eight soldiers could be seated on benches along the interior side walls. Although classified as light APC's these vehicles were often used in place of the medium APC Sd.Kfz.251.

Along with the basic version armed with 2 x 7.92 mm MG 34 or 42, the SPW U304(f) was also equipped with a 3.7 cm Pak 36 as a platoon leader car analogous to the Sd.Kfz.251/10. There was also a version fitted with a mortar 34, radio and command cars and an armoured ambulance based on the Unic P 107.

US troops passing a Unic-APC disabled by a landmine, Normandy 1944. (US Army)

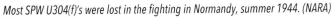

Most SPW U304(f)'s were lost in the fighting in Normandy, summer 1944. (NARA)

Leichter Schützenpanzer-wagen (2-cm Flak) U304 (f)

Type:	Light APC (2-cm flak)
Manufacturer:	Unic, Baukommando Becker
Fighting weight:	5400 kg
Dimensions:	Not available
Motor:	Unic P 39, 4-cylinder petrol engine
Cubic capacity:	3450 ccm
Performance kW/hp:	44/60
Performance weight:	11.1 hp/t
Top speed:	45 km/hr (road)
Fuel capacity:	160 litres
Range:	400 km (road) 160 km (terrain)
Crew:	6
Armament:	1 x 2 cm flak 38
Armour:	Max. 10 mm
Fording depth:	0.8 m

In 1943 and 1944 on the basis of the traction vehicle Unic P 107 two different self-propelled flak cars were made. The first was partially armoured, i.e. only the cooling system and driver's area received a light armour protection. A relatively flat, lightly armoured bodywork was fitted to the loading surface of the Unic P107 and a 2 cm flak 38 mounted at its centre. The second version had a complete armoured superstructure, very similar to the APC 304(f) but a little higher. A 2 cm flak with 360° traverse was located in the crew space. Both versions also had an ammunition trailer.

72 units of the fully armoured variant were built. The overall total of all versions of the APC U304(f) is not known but would not have exceeded a few hundred. Traction vehicles of this kind were operational across Europe, but the APC's practically only in France in units such as the "Fast Brigade West" where they were wiped out in the summer of 1944. All in all, the APC U304(f) failed to prove itself, the engine and electrical systems in particular being very prone to breakdown.

Illustration of the SPW U304 (f) in side profile. (Vincent Bourguignon)

Mittlerer Schützenpanzer- wagen Sd.Kfz. 251/1 Ausf. A and B

Sd.Kfz.251/1 version A against a backdrop of Berlin cathedral and the Old Museum in Berlin, 1941. (BA cc-by-sa 3.0).

In 1937 the Hannover Maschinenbau AG (Hanomag) received a contract from the Army Weapons Department to convert the H kl 6 bogie of its 3-tonne traction vehicle (also known as Sd.Kfz.11) in such a way that an armoured vehicle could be created from it. The armoured superstructure was developed by Büssing-NAG and Deutsche Werke Kiel in the angular form typical for German armoured scout cars of the time. The front armour was 12 mm, the sides and rear 8 mm. As with traction vehicles the front axle of the medium APC had no brakes or drive, being used only for steering. To assist sharp steering movements a track-brake was used. The six pairs of large wheels had torsion bar suspension and were arranged in

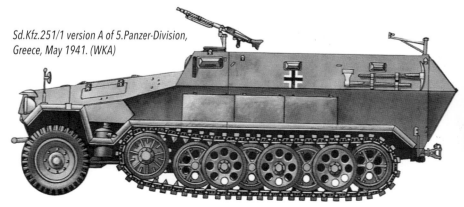

Sd.Kfz.251/1 version A of 5.Panzer-Division, Greece, May 1941. (WKA)

Sd.Kfz.251/1 version B in front of the Berlin city Schloss, 1940. (BA cc-by-sa 3.0)

staggered order with rubber tread. The drive wheels were the leading pair. Greased crawler tracks with rubber cushioning provided a relatively quiet and smooth ride. The water-cooled 100-hp Maybach engine was situated forward, the open crew space was immediately behind it with seats for the driver and co-driver/radio operator, benches along the inner walls seated ten grenadiers. Equipment and ammunition could be stored below these benches. At the rear of the vehicle was a large, double-flap door opening outwards enabling rapid alighting and re-entering. In bad weather the crew space could be protected by a tarpaulin supported by four hoops. The armament of the basic version was two MG 34 (later also MG 42) with 2010 rounds, the rear MG having a "flak swivelling arm" for pointing at aerial targets. If necessary these MG's could be interchanged.

The first pre-series appeared in 1938 and the first series-built models of version A (called a "medium armoured crew transport vehicle") were delivered to the troops in the spring of 1939 receiving their baptism of fire in the Polish campaign. Model A had three observation hatches either side which were dispensed with in the B version appearing shortly afterwards. The B version also had the typical splinter shield and pack lockers between the upper bodywork

Type:	Medium armoured personnel carrier
Manufacturer:	Hanomag, Borgwand
Fighting weight:	8500 kg
Length:	5800 mm
Breadth:	2100 mm
Height:	1750 mm (without MG shield)
Motor:	Maybach HL 42 TURKM, 6-cylinder petrol engine
Cubic capacity:	4170 ccm
Performance kW/hp:	73.6/100
Performance weight:	11.8 hp/t
Top speed:	52.5 km/hr (road)
Fuel capacity:	160 litres
Range:	320 km (road) 180 km (terrain)
Crew:	2 + 10
Armament:	2 x 7.92 mm MG 34
Armour:	6 to 14.5 mm
Fording depth:	0.5 m

and track guard. These improvements were made to version A later. With effect from version B the front armour was increased to 14.5 mm.

Mittlerer Schützenpanzerwagen Sd.Kfz. 251/1 Ausf. C

brought inside and mounts fixed to carry two MPi 38 and their magazines. Besides other changes in detail access to the rear crawler track section was made easier by removing the rear number plate and painting the registation mark on the wheel guard.

The increased demand for the Sd.Kfz.251 led to a large number of concerns being included in the

After the medium APC had passed its trials in the Polish campaign, improvements based on battle experiences began to flow into the Sd.Kfz.251/1 C version built from mid-1940 receiving as its most striking changes a new one-piece front section and typical covers for the air intake pipes beside the motor. The forward fenders on versions A and B were now omitted. In contrast to these versions in which the entrenching tools were carried on the superstructure, these were now lodged above the wheel guard. The seating in version C was adjustable and had a back rest and head rest. The fire extinguisher previously outside the vehicle was

Sd.Kfz.251/1 version C of the SS-Panzer-Grenadier-Division Totenkopf, Eastern Front, summer 1943. (WKA)

The Sd.Kfz.251/1 is easily recognized by the covers for the air intake tubes alongside the motor room and the one-piece front armour. (Tank Museum)

Sd.Kfz.251/1 version C of an unknown unit, Eastern Front, summer 1942. (BA cc-by-sa 3.0)

completion stage. Thus the "H kl 6p" bogie was not only produced by Hanomag and Borgward, but also by Adler, Auto-Union, Skoda, Stöwer and MNH. Armoured superstructures were completed at numerous smaller firms such as Schoeller & Bleckmann, Steinmüller and others. Because not all factories had the same equipment this resulted in some superstructures being riveted instead of welded. The final assembly of the medium APC was handled by Weserhütte at Bad Oeynhausen, Schichau/Elbing and Wumag/Görlitz. Besides the basic version of the Sd.Kfz.251/1 which carried ten grenadiers and two light MGs, another with the same classification carried nine men, two heavy MGs with tripod rests and sighting equipment.

Type:	Medium armoured prsonnel carrier
Manufacturer:	Final assembly at Weserhütte, Schichau and others
Fighting weight:	8500 kg
Length:	5800 mm
Breadth:	2100 mm
Height:	1759 mm (without MG shield)
Motor:	Maybach HL 42 TURKM, 6-cylinder petrol engine
Cubic capacity:	4170 ccm
Performance kW/hp:	73.6/100
Performance weight:	11.8 hp/t
Top speed:	52.5 km/hr (road)
Fuel capacity:	160 litres
Range:	320 km (road 180 km (terrain)
Crew:	2 + 10
Armament:	2 x 7.92 mm MG 34 or MG 42
Armout:	6 to 14.5 mm
Fording depth:	0.5 m

Mittlerer Schützenpanzer-wagen Sd.Kfz. 251/1 Ausf. D

Version D of the medium APC was built from 1943 onwards under pressure for a quick output despite the war conditions while taking into account battle experience. The armoured superstructure was much simplified, and angular plates, such as at the rear and sides, replaced by simple, straight armour sheets. The previously removeable pack lockers were now integrated into the superstructure and instead of observation flaps at the sides version D had only viewing slits. Air intake for the engine was put under the forward armour at the side and the front armour increased to 15 mm. This was the final form of the Sd.Kfz 251 to the war's end. About 16,000 Sd.Kfz.251 of all versions were built, 12,000 of these being version D. The versions A to D of the medium APC were manufactured as numerous variants with different equipment and armament. Officially there were 23 versions and several unofficial sub-variants. Besides those mentioned here a variant (Sd.Kfz. 250/4) was envisaged for towing light artillery,Sd.Kfz. 250/8 was an unarmed ambulance version with stretcher-beds, Sd.Kfz.250/11 an armoured telephone-cable layer and Sd.Kfz.250/19 served as an armoured telephone exchange for field telephones. The APC's of the types Sd.Kfz.251/12 to 15 inclusive were special vehicles built in small numbers for the artillery for measuring, sound ranging and light evaluation. Sd.Kfz.251/18 was an

Medical staff, US 359th Inf.Regt/90th.Inf.Div. awaiting German and US wounded leaving an Sd.Kfz.251/1 version D wearing Red Cross flags, near Chambois, France, August 1944. (NARA).

Type:	Medium armoured personnel carrier
Manufacturer:	Final assembly at Weserhütte, Schichau and others
Fighting weight:	8500 kg
Length:	5980 mm
Breaadth:	2100 mm
Height:	1750 mm (without MG shield)
Motor:	Maybach HL 42 TURKM 6-cylinder petrol engine
Cubic capacity:	4170 ccm
Performance kW/hp:	73.6/100
Performance weight:	11.8 hp/t
Top speed:	52.5 km/h (road)
Fuel capacity:	160 litres
Range:	320 km (road) 180 km (terrain)
Crew:	2 + 10
Armament:	2 x 7.92 mm MG 34 or MG 42
Armour:	6 to 15 mm
Fording depth:	0.5 m

artillery spotter version. It is not certain whether the variant Sd.Kfz.251/23 planned in 1945 with the turret of the Sd.Kfz.250/9 or Sd.Kfz.234/1 ever came about. The medium APC was used as the basis for numerous experimental vehicles, e.g. in 1943 an SP-chassis with an 8.8 cm Pak. Although the Sd.Kfz.251 was on the whole a successful design it had a number of deficiencies. The troops complained especially about its poor cross-country performance compared to fully tracked vehicles and the weak armour. The open-top crew space offered little protection against air attack or artillery fire. The efficient but expensive and complex bogie was another drawback. Nevertheless with the Sd.Kfz.251 the Wehrmacht created a completely new, progressive kind of fighting vehicle. After 1945 the medium APC (with some modifications) was produced for years in Czechoslovakia as the OT-810.

On 14 August 1944 during the Warsaw Uprising, Polish partisans captured this Sd.Kfz.251/1 version D of 5.SS-Panzer-Division Wiking. (NARA)

Mittlerer Schützenpanzer-wagen (Wurfrahmen 40) Sd.Kfz. 251/1

Type:	Medium armoured personnel carrier
Fighting weight:	9 tonnes
Height:	5800 mm
Breadth:	2100 mm (without launch frames)
Height:	1750 mm (without MG shield)
Motor:	Maybach HL 42 TURKM 6-cylinder petrol engine
Cubic capacity:	4170 ccm
Performance kW/hp:	73.6/100
Performance weight:	11.1 hp/t
Top speed:	52.5 km/hr (road)
Fuel capacity:	160 litres
Range:	320 km/hr (road) 180 km/hr (terrain)
Armament:	2 x 7.92 mm MG 34 or MG 42, six launch frames 40 for 28-cm or 32-cm rockets
Armour:	6 to 14.5 mm
Fording depth:	0.5 m

Based on the already available Sd.Kfz.251/1, Gast KG in Berlin produced a spectacular version of the APC as a rocket launcher. This variant had a metal framework fitted either side of the bodywork to hold three so-called "Wurfrahmen 40" (launch frame 40) which could be hand-adjusted for elevations between +5° and +40° to fire rocket-propelled warhead of 28 and 32 cm calibre. The 28 cm weapon was an explosive and had a maximum range of 1900 metres. The 32 cm weapon was filled with inflammable oil and had a range of up to 2200 metres. The vehicle had to be pointed in the direction of fire. Although not very accurate, these weapons had a devastating effect at ground level. Usually one oil- and five explosive rockets were carried and could be fired one after another with a slight delay between each. APC's of this kind were usually operated by pioneer units and were dubbed Stuka zu Fuss (walking Stukas) by the men.

Opposite Top: Sd.Kfz.251, 24.Panzer-Division with heavy launch frames 40, Voronezh area, June/July 1942. (NARA)

Opposite Bottom: Wurfgranate rocket after being fired from an Sd.Kfz.251. (NARA)

Sd.Kfz.251/1 version A showing the lateral launch frame mountings. (BA cc-by-sa 3.0)

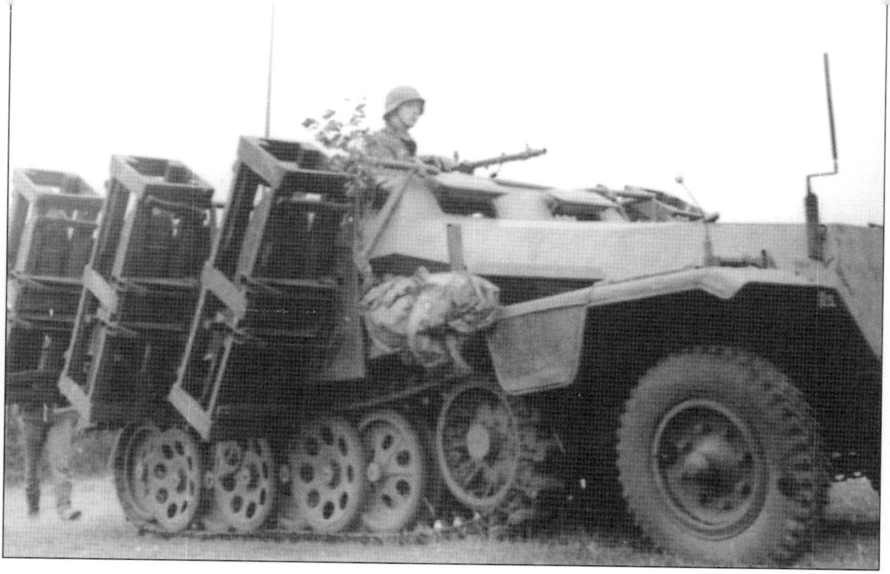

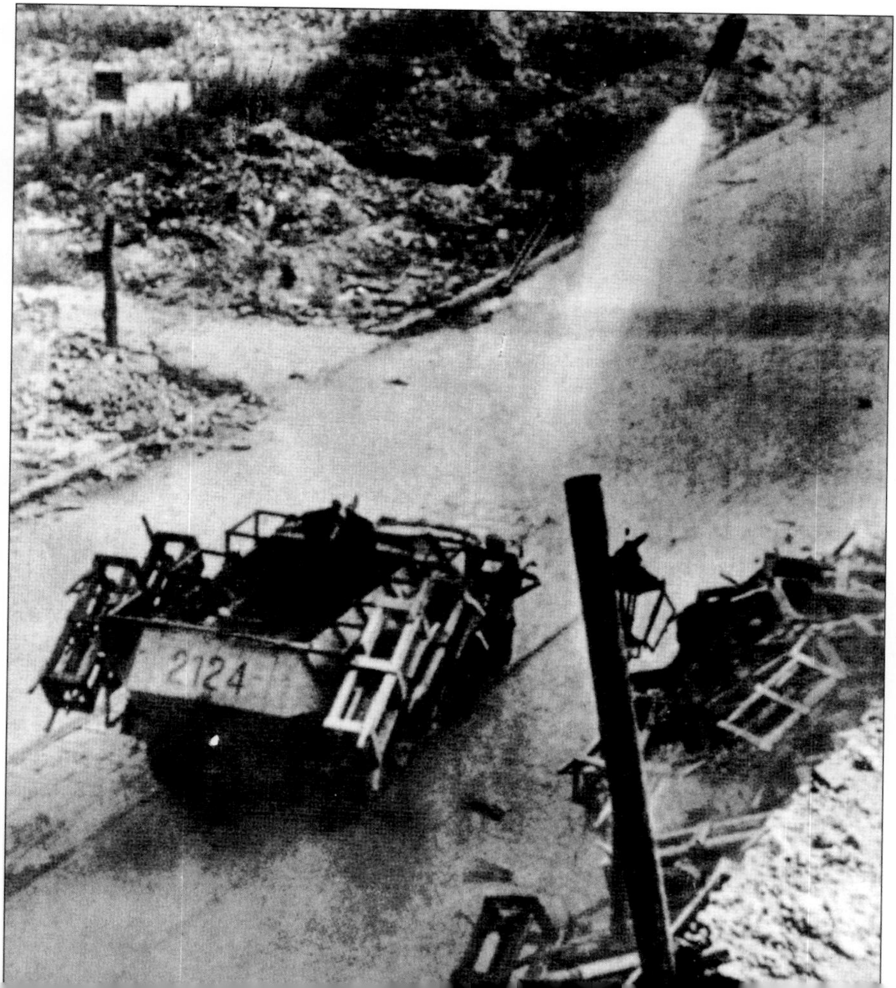
2124

Mittlerer Schützenpanzer-wagen (sGrW.34) Sd.Kfz. 251/2

Type:	Medium armoured personnel carrier
Fighting weight:	8640 kg
Length:	5800 mm
Breadth:	2100 mm
Height:	1750 mm (without MG shield)
Motor:	Maybach HL 42 TURKM 6-cylinder petro engine
Cubic capacity:	4170 ccm
Performance kW/hp:	73.6/100
Performance weight:	11.6 hp/t
Top speed:	52.5 km/hr (road)
Fuel capacity:	160 litres
Range:	320 km (road) 180 km (terrain)
Crew:	2 + 6
Armament:	1 x 7.92 mm MG 34 or MG 42, 1 x 8 cm mortar 34
Armour:	6 to 14.5 mm
Fording depth:	0.5 m

The Sd.Kfz.251/2 built from 1940 was used to carry a heavy mortar (sGrW) and its six-man crew. The 8-cm Granatwerfer 34 could be fired from within the APC, but was normally unloaded and set up beyond it. For this purpose the base plate of the mortar was carried on the front plating over the motor. In order to create more room for ammunition, some of the bench seating in the crew space was removed, enabling the APC to carry sixty-six 8-cm mortar bombs each weighing 3.5 kg. These could be fired a maximum range of 2400 metres. Other armament was an MG 34 or 42 with 2010 rounds. In common with all medium APC's this version also had a radiophone "f" with a 2-cm high rod aerial. In the course of the war they were often replaced by light APCs for the same purpose but remained operational until the capitulation.

Sd.Kfz. 251/2, unknown unit, Eastern Front, winter 1942/1943. (Vincent Bourguignon)

Mittlerer Funkpanzerwagen Sd.Kfz. 251/3

There were not less than nine versions of the medium armoured radio car, distinguished from each other by the radio equipment carried. Accordingly, depending on what radio was installed, the Sd.Kfz.251/3 served as an armoured radio centre for liaison to divisional or regimental command posts, as a command vehicle (especially for panzer and panzer-grenadier units), as a batallion command car or as an APC for flight liaison officers and also for general communications. Medium armoured radio cars would be attached to the signals section of a unit to replace the radio-equipped armoured scout cars of the early war years. From 1942 the frame aerial was often replaced by a less obvious umbrella aerial. Armament consisted of one to two MG 34 or MG 42 with 2010 rounds as with the basic version. Cars of this type were built from 1940 and remained operational in all theatres of war until 1945.

Type:	Medium armoured radio car
Fighting weight:	8500 kg
Length:	5800 mm
Breadth:	2100 mm
Height:	1750 mm (without MG shield or aerial)
Motor:	Maybach HL 42 TURKM, 6-cylinder petrol engine
Cubic capacity:	4170 ccm
Performance kW/hp:	73.6/100
Performance weight:	11.8 hp/t
Top speed:	52.5 km(hr (road)
Fuel capacity:	160 litres
Range:	320 km (road) 180 km (terrain)
Crew:	2 + 5
Armament:	1 or 2 x 7.92 mm MG 34 or MG 42
Armour:	6 to 14.5 mm
Fording depth:	0.5 m

Sd.Kfz 251/3 version B, 21.Panzer-Division, Libya 1941. (Vincent Bourguignon)

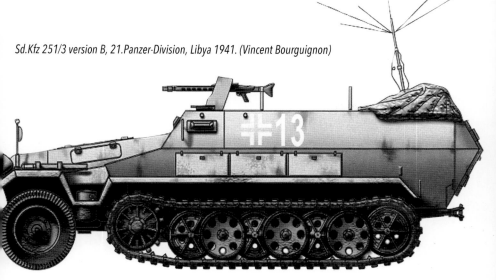

Mittlerer Pioniergeräte- wagen Sd.Kfz. 251/5 und 251/7

Type:	Medium pioneer equipment car, (data for Sd.Kfz.251/7 version D)
Manufacturer:	Final assembly at Weserhütte, Schichau and others
Fighting weight:	8850 kg
Length:	5980 mm
Breadth:	2100 mm (without bridge parts)
Height:	1750 mm (without MG shield)
Motor:	Maybach HL 42 TURKM, 6-cylinder petrol engine
Cubic capacity:	4170 ccm
Performance kW/hp:	73.6/100
Performance weight:	11.3 hp/t
Top speed:	52.5 km/hr (road)
Fuel capacity:	160 litres
Range:	320 km (road) 180 km (terrain)
Crew:	2+5
Armament:	2 x 7.92 mm MG 34 or MG 42, 1 x 7.92 mm PzB 39
Armour:	6 to 16 mm
Fording depth:	0.5 m

These two models of the medium APC were special versions for the panzer-pioneers. In order to create more space in the interior of the vehicle for pioneers' equipment such as tools, mines or explosives, one of the two longitudinal benches of the basic version was discarded. Sd.Kfz.251/5 also had a comprehensive radio installation and was used mainly by platoon leaders, Sd.Kfz.251/7 carried at the side of the armoured superstructure a framework for parts of an assault bridge. Besides the usual MGs a 7.92 mm Panzerbüchse 39 anti-tank gun and 40 rounds were carried.

Pioneer equipment cars were built on the basis of all versions of the Sd.Kfz.251 and show clearly how the needs of the panzer and panzergrenadier divisions had been well thought-through initially in order to make every division into an entity in a position to act independently without having to call upon outside support. A concept which contributed substantially to the Wehrmacht's initial successes.

Sd.Kfz.251/7 version D at the Panzer Museum, Munster. Note the bridge sections on the sides of the superstructure. (Ikeda Shinobu).

Mittlerer Kommando-panzerwagen Sd.Kfz. 251/6

Type:	Medium armoured command car
Manufacturer:	Final assembly at Weserhütte, Schichau and others
Fighting weight:	8800 kg
Length:	5800 mm
Breadth:	2100 mm
Height:	2700 mm (with frame aerial)
Motor:	Maybach HL 42 TURKM 6-cylinder petrol engine
Cubic capacity:	4170 ccm
Performance kW/hp:	73.6/100
Performance weight:	11.4 hp/t
Top speed:	52.5 km/hr (road)
Fuel capacity:	160 litres
Range:	320 km (road) 180 km (terrain)
Crew:	2 + 6
Armament:	1 x 7.92 mm MG 34
Armour:	6 to 14.5 mm
Fording depth:	0.5 m

The Sd.Kfz. 251/6 was a command version of the medium APC for senior officers, e.g. General Guderian used one such during the campaign in France. It had comprehensive appointments by way of map tables, encryption machines and a powerful radio installation (FuG 11 and FuG Tr 100mw). The frame aerial for the latter was often replaced by an umbrella aerial during the course of the war. The crew space, open to the elements at the top, could be covered over by a tent-like tarpaulin if required. An MG with 1100 rounds was the only armament. Armoured command vehicles of this sort were used mainly by Army Group leaders and divisional commanders and were therefore not produced in any great numbers. Furthermore the deteriorating war economy found such costly vehicles a luxury from 1943 and for this reason the type was discontinued that year.

Sd.Kfz.251/6 version A. (WKA)

Mittlerer Schützenpanzer-wagen (7.5 cm StuK 37) Sd.Kfz. 251/9

Type:	Medium armoured personnel carrier
Manufacturer:	Final assembly at Weserhütte, Schichau and others
Fighting weight:	8800 kg
Length:	5980 mm
Breadth:	2100 mm
Height:	2070 mm
Motor:	Maybach HL 42 TRKM 6-cylinder petrol engine
Cubic capacity:	4170 ccm
Performance kW/hp:	73.6/100
Performance weight:	11.4 hp/t
Top speed:	52.5 km/hr (road)
Fuel capacity:	160 litres
Range:	320 km (road) 180 km (terrain)
Crew:	3
Armament:	1 x 75 mm StuK 37, 1 x 7.92 mm MG 34 or MG 42
Armour:	6 to 15 mm
Fording depth:	0.5 m

This version of the Sd.Kfz.251 (also known by the troops as *Stummel*-"stumpy") was one of the many emergency solutions and improvisations to which the Wehrmacht was forced to make during the course of the war. Because the panzers or assault guns intended to give fire support to panzergrenadier units were often simply not available, the units were provided with the Sd.Kfz.251/9, a vehicle built expressly to fill the breach.

At the end of March 1942 Büssing-NAG received the contract to develop a version of the APC with 7.5 cm StuK 37 L/24. The first prototypes were tested on the Eastern Front in June 1942 and series production began shortly after. The short-barrelled StuK, for which 52 rounds were carried, had a low muzzle velocity and was situated close to the driver's right side. The traverse was only 12° to either side. The gun could be moved in the vertical register between -10° depression and +12° elevation. A choice of AP, smoke and hollow charge shells were available for use against enemy armour. Later vehicles had higher lateral sides to protect the crew. A total of 1141 of these cars were built, 582 were still on the books on 1 March 1945.

Sd.Kfz.251/9 version D exhibited at the Panzer Museum, Munster. (Ikeda Shinobu)

Mittlerer Schützenpanzerwagen (3.7-cm-Pak 36) Sd.Kfz. 251/10

Type:	Medium armoured personnel carrier
Manufacturer:	Final assembly at Weserhütte, Schichau and others
Fighting weight:	8500 kg
Length:	5800 mm
Breadth:	2100 mm
Height:	1750 mm (without Pak)
Motor:	Maybach HL 42 TURKM, 6-cylinder petrol engine
Cubic capacity:	4170 ccm
Performance kW/hp:	73.6/100
Performance weight:	11.8 hp/t
Top speed:	52.5 km/hr (road)
Fuel capacity:	160 litres
Range:	320 km (road) 180 km (terrain)
Crew:	2 + 4
Armament:	**1** x 3.7 cm Pak 36 L/45, 1 x 7.92 mm MG 34 or MG 42, 1 x 7.92 mm PzB 39
Armour:	6 to 14.5 mm
Fording depth:	0.5 m

The Sd.Kfz.251/10 was the first APC to be armed with a cannon and was received by the troops from 1940. The 3.7 cm Pak 36 with 5 mm shield but less wheels and chassis tail was mounted in place of the forward MG. Later a lower and broader shield which stretched across the breadth of the vehicle replaced the earlier one, but often only the left part was mounted to give the gunlayer protection at the telescopic sight. The Sd.Kfz.251/10 was conceived as a company commander's car and to provide supporting fire to other APCs of the platoon. The effectiveness of the Pak 36 against armoured targets fell off rapidly after 1940, but it remained very useful against MG nests, lightly armoured targets or field fortifications. Munition supply for the Pak 36 was 162 shells.

Besides an MG 34 or 42 (2010 rounds) the APC was fitted with a Panzerbüschse 39 (7.92 mm calibre) anti-tank weapon with 40 rounds. Production of the Sd.Kfz.251/10 ended in 1943 but cars of this type remained operational until the war's end.

Sd.Kfz.251/10 of an unknown unit, northern sector of Eastern Front, beginning 1943. (Vincent Bourguignon)

Mittlerer Flammpanzer-wagen Sd.Kfz. 251/16

The armament of the Sd.Kfz.251/16 consisted of one to two 7.92 mm MG 34 or MG 42 and two 14 mm flamethrower tubes located behind armour shielding at the sides of the upper superstructure. The tubes could be sivelled up to 90° to the side and elevated +40°. Initially 7 mm tubes were installed instead of 14 mm on some vehicles. A supply source in the crew space was enough for eighty pulses of burning oil over a range of about 35 metres. A portable device was also carried which had a range of 10 metres. The flamethrower armoured vehicles attached to panzer pioneer units were intended primarily for use against infantry in well constructed positions and bunkers.

Type:	Medium armoured flamethrower
Manufacturer:	Final assembly at Weserhütte, Schichau and others
Fighting weight:	8620 kg
Length:	5800 mm
Breadth:	2100 mm
Height:	2010 mm (with flame tubes)
Motor:	Maybach HL 42 TURKM 6-cylinder petrol engine
Cubic capacity:	4170 ccm
Performance kW/hp:	73.6/100
Performance weight:	11.6 hp/t
Top speed:	52.5 km/hr (road)
Fuel capacity:	160 litres
Range:	320 km (road) 180 km (terrain)
Crew:	3 to 5
Armament:	1 to 2 x 7.2 mm MG 34 or Mg 42, 2 x 14 mm flame tubes, 1 x 7 mm flame tube
Armour:	6 to 14.5 mm
Fording depth:	0.5 m

Sd.Kfz.251/16 version A. (Vincent Bourguignon)

Mittlerer Schützen-panzerwagen (2-cm-Flak 38) Sd.Kfz. 251/17

Type:	Medium armoured personnel carrier
Manufacturer:	Final assembly at Weserhütte, Schichau and others
Fighting weight:	8900 kg
Length:	5800 kg
Breadth:	2100 mm
Height:	2250 mm
Motor:	Maybach HL 42 TRKM 6-cylinder petrol engine
Cubic capacity:	4170 ccm
Performance kW/hp:	73.6/100
Performance weight:	11.4 hp/t
Top speed	52.5 km/hr (road)
Fuel capacity:	160 litres
Range:	320 km (road) 180 km (terrain)
Crew:	5 to 7
Armament:	1 x 2 cm Flak 38 with 600 rounds, 1 x MG 34 or MG 42
Armour:	6 to 14.5 mm
Fording depth:	0.5 m

The Sd.Kfz.251/17 was a flak version of the medium APC of which there were four versions. In the first version which appeared in 1942, the Flak 38 was situated in the open crew space. The narrow open-top area severely limited the traverse so that in the so-called "Luftwaffe version" the rear armoured superstructure was broadened and hinged sides fitted. This gave the flak a 360° traverse but the gun crew now had scanty protection. Another version had an unarmoured platform with drop-sides which offered the crew even less protection. As all three versions were unsatisfactory, from September 1944 a new variant in limited numbers was produced based on the bogie of version D with the 2-cm flak on a swivelling chassis providing the maximum traverse and elevation while the crew remained inside the crew space.

The so-called "Luftwaffe version" of the Sd.Kfz.251/17 with elongated box upper part, the side walls having drop-hinges. (WKA)

Schützenpanzer-wagen Sd.Kfz. 251/20 (Uhu)

Type:	Medium armoured personnel carrier
Manufacturer:	Final assembly at Weserhütte, Schichau and others
Fighting weight:	8500 kg
Length:	5980 mm
Breadth:	1750 mm (without search light)
Height:	2100 mm
Motor:	Maybach HL 42 TRKM 6-cylinder petrol engine
Cubic capacity:	4170 ccm
Performance kW/hp:	73.6/100
Performance weight:	11.8 hp/t
Top speed:	52.5 km/hr (road)
Fuel capacity:	160 litres
Range:	320 km (road) 180 km (terrain)
Crew:	3 to 4
Armament:	None
Armour:	6 to 15 mm
Fording depth:	0.5 m

Because of Allied air supremacy, movements by armoured Wehrnacht units in 1944 in daylight were scarcely possible. Therefore it was decided to employ infra-red (IR) for night driving and night aiming equipment. By the end of 1944 the IR system Puma for the PzKpfw.V *Panther*, and the system Falke for the APC Sd.Kfz.251 were ready. This equipment consisted of a small IR-searchlight and an aiming device with image changer. Because the range of both was too short, a vehicle Sd.Kfz. 251/20 with 60-cm infra-red searchlight Beob.Ger.12/51 and known as Uhu ("owl") lit up the nocturnal battlefield. For the purpose a 60-cm anti-aircraft searchlight (with carbon arc lamp and infra-red filter) and an aiming device with image changer was mounted in the crew space of a medium APC. The searchlight could be traversed 360°. The current source to work the Beob.Ger.12/51 was also located in the crew space. The maximum range of the infra-red searchlight was 1500 metres. Only a handful of these vehicles were operational before the war's end but according to reports they achieved astounding results in combination with correspondingly equipped *Panthers*.

An Sd.Kfz.251/20 at a US Army vehicle assembly point in 1945. (US Army)

Apart from the large searchlight in the crew room, the driver of the Sd.Kfz.251/20 also had a small infra-red searchlight and an image changing apparatus. (US Army)

Flakschützen-panzerwagen Sd.Kfz. 251/21

Type:	Flak armoured personnel carrier
Manufacturer:	Final assembly at Weserhütte, Schichau and others
Fighting weight:	8620 kg
Length:	5980 mm
Breadth:	2100 mm
Height:	2100 mm
Motor:	Maybach HL 42 TRKM, 6-cylinder petrol engine
Cubic capacity:	4170 ccm
Performance kW/hp:	73.6/100
Performance weight:	11.6 hp/t
Top speed:	52.5 km/hr (road)
Fuel capacity:	160 litres
Range:	320 km/hr (road) 180 km/hr (terrain)
Crew:	4 to 6
Armament:	3 x 1.5 cm MG 151/15 or 2 cm MG 151/20 with 2000 rounds
Armour:	6 to 15 mm
Fording depth:	0.5 m

In order to improve the defence against air attack, the Army Weapons Department sought ideas for a multi-barrelled weapon to put on the chassis of the Sd.Kfz.251. The 2-cm quadruple flak was too large and heavy and the vehicle would not carry an adequate stock of ammunition for it. The solution was found eventually in the shape of the on-board aircraft cannon MG 151/15 or MG 151/20 (calibre 15-20 mm) mounted on a hand-operated triple chassis. The side walls of the armoured superstructure were raised and the chassis protected by a shield such that when lowered it looked almost like a turret. Because the MG 151 ammunition took up less space than the Flak 38, 2000 rounds could now be carried. Although the vehicle had only a hand-held optical rangefinder and no other ranging devices, it was successful for its high cadence. Between August 1944 and the war's end a total of 387 were built.

Sd.Kfz.251/21 (WKA)

Mittlerer Schützenpanzer-wagen (7.5-cm Pak 40) Sd.Kfz. 251/22

Type:	Medium armoured personnel carrier
Manufacturer:	Final assembly at Weserhütte, Schichau and others
Fighting weight:	9900 kg
Length:	5980 mm
Breadth:	2100 mm
Height:	1750 mm (without Pak)
Motor:	Maybach HL 42 TRKM 6-cylinder petrol engine
Cubic capacity:	4170 ccm
Performance kW/hp:	73.6/100
Performance weight:	10.1 hp/t
Top speed:	50 km/hr (road)
Fuel capacity:	160 litres
Range:	320 km (road) 180 km (terrain)
Crew:	4
Armament:	1 x 7.5 cm Pak 40 L/46, 1 x 7.92 mm MG 42
Armour:	6 to 15 mm
Fording depth:	0.5 m

The 7.5 cm Pak 40 anti-tank gun crews found the gun so heavy as to be almost immovable. In order to give the weapon mobility a self-propelled tank hunter was designed on the basis of the Sd.Kfz.234 and also 251. As with the Sd.Kfz.234/4 the whole upper part of the Pak, plus shield but less wheels and chassis tail, was placed on a pivot in the crew room. At the end of November 1944, Hitler himself urged that the supply of the vehicle to the frontline troops should have the highest priority. Once examples of the vehicles reached the units, they became responsible themselves to make the changes. All new Pak 40 to be installed in the Sd.Kfz.251 were to have the shield reduced.

Although only an emergency solution, the vehicles were very effective, even though the gun overburdened them.

Replica of an Sd.Kfz. 251/22 from a Czech APC OT-810. (Martin Spurny)

Mittlerer Schützenpanzer-wagen (Prototyp) HKp 606

Type:	Medium armoured personnel carrier
Manufacturer:	DEMAG
Fighting weight:	7 tonnes
Length:	4850 mm
Breadth:	1980 mm
Height:	1850 mm (upper edge bodywork)
Motor:	Maybach HL 50 6-cylinder petrol engine
Cubic capacity:	5000 ccm
Performance kW/hp:	132/180
Performance weight:	25.7 hp/t
Top speed:	70 km/hr (road)
Fuel capacity and range:	Not available
Crew:	2 + 8
Armament: planned	2 x 7.92 mm MG 42
Armour:	6 to 14.5 mm
Fording depth:	Not available

Borgward (previously Hansa-Lloyd-Goliath), DEMAG and Hanomag began work in 1939 on a successor model for the 1-tonne and 3-tonne traction vehicles and the Sd.Kfz.250 and 251 based on them. The experimental vehicles KHp 602 (DEMAG) and HKp603 (Hanomag) were basically modified Sd.Kfz.251 with a new Maybach HL 45 Z engine, new drive system and an altered front. These did not enter series production, and neither did the DEMAG prototype HKp 605 in 1941-1942. Model HKp 606 also built by DEMAG in 1942 concluded the development and was to have replaced all other APCs. Amongst the features of the new APC were a simplified armoured superstructure and a more powerful Maybach 180-hp HL 50 engine with Maybach OLVAR pre-select gears. Ultimately the development was halted and preparations for series production abandoned once the Army Weapons Department and Guderian formed the opinion that it would lose too much time. Instead new, simplified versions of the light and medium APCs were forced through, their armoured superstructures reminders of the HKp 606 prototypes.

SPW HKp 606. (WKA)

Mittlerer Schützenpanzerw agen S307 (f)

As from 1933 SOMUA built for the French armed forces 2,543 half-track towing vehicles of the MCG 5 model, having tracks of the Kégresse-Hinstin type, reinforced endless rubber tracks and leaf suspension. Towing tractors of this kind were given various special types of superstructure and used in a number of roles, principally as artillery tugs and transporters. A number of these vehicles fell into Wehrmacht hands in the summer of 1940. After overhaul at the manufacturers they were then used by German forces for the originally intended purpose under designation Zgkw S307(f).

In 1943/44 Baukommando Becker working in collaboration with French industry created a series of different armoured vehicles on the basis of the SOMUA-MCG bogie. Additionally the original upperworks were replaced by armoured

Although the 7.5 cm Pak was too heavy for the APC S307(f) bogie, the solution at least enabled the heavy Pak to be relatively mobile. (Vincent Bourguignon)

superstructures fairly similar in form to the medium APC Sd.Kfz.251. As a result, in 1943 an unknown number of APCs and armoured pioneer cars were turned out. 48 armoured munitions carriers, 36 light multiple launchers in two slightly differing variants and a small number of vehicles with the 15 cm

An APC S307(f) with 7.5 cm Pak captured by US forces in Normandy, summer 1944. (NARA)

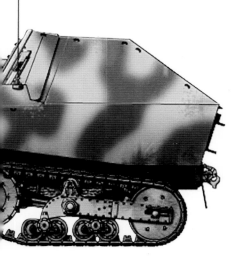

Type:	Medium armoured personnel carrier
Manufacturer:	SOMUA/France; Baukommando Becker
Fighting weight:	7300 kg (data for 7.5 cm Pak 40 on S 307(f)
Length, breadth and height:	Not available
Motor:	SOMUA 4-cylinder petrol engine
Cubic capacity:	4712 ccm
Performance kW/hp:	44/60
Performance weight:	8.2 hp/t
Top speed:	36 km/hr (road)
Fuel capacity:	80 litres
Range:	170 km (road)
Crew:	4
Armament:	1 x 7.5 cm Pak 40 L/46
Armour:	Max 10 mm
Fording depth:	Not available

Panzerwerfer 42 rocket were also built. The best known conversions of this type are undoubtedly the tank hunters built in 1944 on the MCG bogie and designated "7.5 cm Pak 40(Sf) on medium APC S307(f)." The Pak was mounted at the rear of the vehicle in the open-top crew area and had only a limited traverse. 72 units of this improvised tank hunter were produced for the "Rapid Brigade West".

As with the APC U304(f) these conversions failed to satisfy. Constant engine and electrical breakdowns and, because of the war situation, the poor quality of the armour, substantially reduced their fighting value.

A wrecked "leichter Reihenwerfer" (light multiple launcher) on S307(f) bogie, Normandy, summer 1944. (NARA)

Heavy multiple launcher on APC 303(f): (Vincent Bourguignon)

8-cm multiple launcher on APC S303(f). (Vincent Bourguignon)

Mittlerer Schützenpanzerw agen S303 (f)

The SOMUA type MCL was a heavier and more efficient version of the MCG and used primarily by the French Army for hauling heavy guns and retrieving damaged vehicles. Between 1933 and 1940, SOMUA built a total of 796 tugs of this type. Those captured by the Wehrmacht were overhauled and then used against as towing tractors for artillery under the designation ZgKw 303(f). In 1943 based on these vehicles a small number of medium APCs and six rocket launchers were constructed (8-cm multiple launcher on APC S303(f).) These vehicles carried an 8-cm launcher for 48 rockets at the rear. The launcher had a 360° traverse and could be elevated up to +45°. Rockets and launchers were copies of the well known Soviet "Stalin organ" and "Katyusha". Maximum range was 5,300 metres. Sixteen "heavy multiple launcher" vehicles were completed from 1943. At the rear of the superstructure these vehicles had a revolving platform on which twenty rocket launchers were arrayed in two rows. They were of French origin (Model Brandt 278(f), calibre 81.4 mm) and could be fired singly or in salvoes. The barrels could be raised to a maximum 90°.

The "light multiple launcher" on the S307(f) bogie was similar in structure but had only sixteen

launchers. As with other armoured conversions on the basis of French half-tracks the vehicles were deployed only with the "Rapid Brigade West" (later part of 21.Panzer-Division) in France. These improvisations were not found very satisfactory.

Type:	Medium armoured personnel carrier
Manufacturer:	SOMUA/France, Baukomman do Becker
Fighting weight:	6850 kg (data for 8 cm multiple launcher on APC S303(f)
Length, breadth and height and fording depth:	Not available
Motor:	SOMUA, type 23 4-cylinder petrol engine
Cubic capacity:	6232 ccm
Performance kW/hp:	59/80
Performance weight:	11.6 hp/t
Top speed:	34 km/hr (road)
Fuel capacity:	120 litres
Range:	100 km (road)
Crew:	4
Armament:	Launcher for 1 x 48 rockets
Armour: max.	10 mm

Gepanzerter Zugkraftwagen 8 t Sd.Kfz.7

Type:	Armoured tractor
Manufacturer:	Krauss-Maffei, Daimler-Benz, Büssing-NAG, Sauerer, Borgward
Fighting weight:	11,550 kg
Length:	6850 mm
Breadth:	2400 mm
Height:	2660 mm
Motor:	Maybach HL 62 TUK 6-cylinder petrol engine
Cubic capacity:	6191 ccm
Performance kW/hp:	103/140
Performance weight:	12.1 hp/t
Top speed:	50 km/hr (road)
Fuel capacity:	213 litres, later 203 litres
Range:	250 km (road) 135 km (terrain)
Crew:	12
Armament:	None
Armour: max.	8 mm
Fording depth:	0.65 m

The 8-tonne tractor (Sd.Kfz.7) was developed by Krauss-Maffei from 1934 and used primarily by the Luftwaffe to pull their 8.8-cm flak and 15-cm field howitzer. From 1937 after changes to the chassis unit and engine the final version (KM m 11) was produced, 12,000 of which were delivered to the troops until 1945. Krauss-Maffei itself built 6,120 Sd.Kfz.7, the remainder came from Daimler-Benz, Büssing-NAG, Sauerer and Borgward under licence. In 1938 and 1939 a small number of partly armoured 8-tonners were produced to tow the 8.8-cm flak deployed principally against ground targets. Alongside Army flak, Army anti-tank units were also equipped with these vehicles. The armour was limited to the forward engine area, cooling system and the driver and crew spaces.

In 1944 a number of Sd.Kfz.7 were converted into mobile fire control positions for the V-2 rocket (Sd.Kfz.7/9). These vehicles served not only as control centres for the rocket launch but also towed the launch platforms. They had an armoured superstruture at the rear to protect the crew against exhaust gases released by the rocket at launch.

Between 1939 and 1940, eight-tonne towing tractors were fitted with light armour and used to pull 8.8 cm flak guns. (WKA)

Panzer-Selbstfahrlafette 7.5-cm auf Fahrgestell m.Zgkw. 5t

In 1934 Rheinmetall and Büssing-NAG cooperated in the development of an armoured self-propelled half-track bogie. In the years subsequent this work resulted in a 0-series of three prototypes all armed with a 7.5-cm cannon L/40.8 located in an open-top turret rotatable through 360°. The gun had a muzzle velocity of 685 m/sec and was designed to be the equivalent of anything at that time existing in tank construction. Apparently it was decided against having an MG. All three versions were based on modified bogies of the Büssing-NAG 5-tonne towing tractor (Sd.Kfz.6), although in contrast to it the engine was located at the rear.

The cars of the first and second prototypes (bogie BN 10 H) had only minor differences between them. Both had a narrow, elongated body, the side armour angled away from the direction of incoming fire. Version I however had one roller wheel less than the

II, while the respective turrets and cannon were not identical. The cannon of version II was flatter with a different design of shield and was fitted with a muzzle brake. Both vehicles were around 6 tonnes in weight and could make 60 km/hr on road surfaces. Armour was from 8 to 20 mm. Three cars were built altogether, but no series run ensued.

Version III (bogie HKp 902) was much broader at the front and the side walls of the bodywork were no longer slanted. A single metal guard protected the forward pair of running wheels and the half-track. Although this prototype had armour only 6 to 10 mm thick, it was 11 tonnes heavier. The engine was a 150-hp Maybach HL 45 providing a top speed of 50 km/hr. Two of these vehicles were deployed in North Africa, one being hit and wrecked, the other captured. The subsequent whereabouts of the wreck are unknown. Another was converted in 1941 into a 5-cm SP-flak and a fourth turned up shortly before the war's end as a fire control panzer for the V-2 rocket. There was also no series run for version III.

Based on a modified bogie of the 3-tonne towing tractor (HL kl 3 (H)), in 1936 Hansa-Lloyd-Goliath manufactured a very similar one-off prototype SP-chassis having a 3.7-cm Pak L/70 in a rotating turret and two MG 34's.

Prototype of the armoured SP-chassis II 7.5 cm on m.Zgkw.5 bogie. (Vincent Bourguignon)

Left: A car of the third version of the 7.5 cm half-track SP chassis (bogie HKp 902) was deployed in North Africa and damaged there. (US Army)

Centre: Prototype of the armoured SP-chassis II 7.5 cm seen from ahead. (US Army, Patton Collection)

Type:	Tank hunter on SP-chassis (data for armoured SP II 7.5 cm on 5-tonne tractor bogie)
Manufacturer:	Rheinmetall, Büssing-NAG
Fighting weight:	6 tonnes
Length, breadth, height, fuel capacity and range:	Not available
Motor:	Maybach NL 38, 6-cylinder petrol engine
Cubic capacity:	3791 ccm
Performance kW/hp:	73.6/100
Performance weight:	16.7 hp/t
Top speed:	60 km/hr (road)
Crew:	4
Armament:	1 x 7.5 cm cannon L/40.8
Armour:	8 to 20 mm
Fording depth:	0.6 m

Fire-direction panzer on HKp 902 bogie seen at the Rocket Testing range, Peenemünde, 1944. (BA cc-by-sa 3.0)

2-cm-Flak auf Fahrgestell Zugkraftwagen 1-t Sd.Kfz. 10/4

The 1-tonne tractor of the D 7 bogie (Sd.Kfz.10) was developed by DEMAG into the Sd.Kfz.10/4 used primarily as a traction vehicle for light guns. DEMAG, Adler, Büssing-NAG, Saurer, Phänomen, MIAG and MNH built around 25,000 type D 7 bogies between 1937 and 1945 of which a good 7,500 were intended for the Sd.Kfz.250/252/253 series. The Sd.Kfz.10/4 was one of the first Army flak vehicles and went into series production in 1939. A 2-cm flak 30 with 360° traverse but without protective shield was mounted on the loading surface of the ZgKw. The lateral walls to the rear were hinged and dropped to allow the crew more room to manouevre. At the sides eight boxes each containing two magazines of 20 rounds were fitted, a total of 280 rounds being carried. Any additional ammunition and equipment was carried in a single-axled trailer. As a result of experience gained in the French campaign the 2-cm guns were given a shield.

(Top) Sd.Kfz.10/4 of a Luftwaffe unit in action against ground targets, Eastern Front, November/December 1943. (BA cc-by-sa 3.0)

(Right) Improvised armour added by the crew to a 2-cm flak on a 1-tonne tractor. (WKA)

Type:	Flak gun on SP-chassis
Manufacturer:	DEMAG and numerous other firms
Fighting weight:	5500 kg
Length:	4750 mm
Breadth:	1930 mm
Height:	2 metres
Motor:	Maybach HL 42 TRKM, 6-cylinder petrol engine
Cubic capacity:	4199 ccm
Performance kW/hp:	73.6/100
Performance weight:	18.2 hp/t
Top speed:	65 km/hr (road)
Fuel capacity:	110 litres
Range:	280 km (road) 150 km (terrain)
Crew:	7 to 8
Armament:	1 x 2 cm Flak 30 or 38 L/65
Armour:	Max. 8 mm
Fording depth:	0.7 m

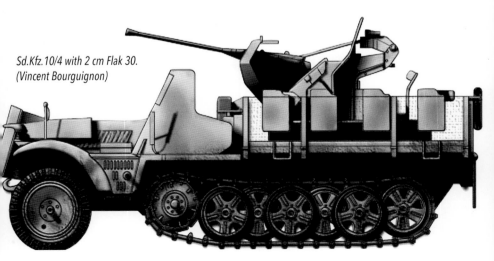

Sd.Kfz.10/4 with 2 cm Flak 30.
(Vincent Bourguignon)

5-cm-Pak auf Fahrgestell Zugkraftwagen 1t Sd.Kfz. 10

Type:	Anti-tank gun on SP chassis
Manufacturer:	DEMAG and numerous other firms
Fighting weight:	5500 kg
Length:	4750 mm
Breadth:	1840 mm
Height:	Not available
Motor:	Maybach HL 42 TRKM 6-cylinder petrol engine
Cubic capacity:	4199 ccm
Performance kW/hp:	73.6/100
Performance weight:	18.2 hp/t
Top speed:	65 km/hr (road)
Fuel capacity:	110 litres
Range:	280 km (road) 150 km (terrain)
Crew:	6
Armament:	1 x 5 cm Pak 38 L/60
Armour:	Max 8 mm
Fording depth:	0.7 m

The main purpose of the Sd.Kfz.10 was to tow light guns such as the 3.7-cm Pak 36 or the 5-cm Pak 38. In order to make these weapons more mobile and get them ready to fire quicker, on their own initiative troops often turned the ZgKw into improvised tank hunters with the Pak mounted on the loading surface. Frequently the entire Pak was lifted up complete with wheels and chassis-tail. These conversions were sometimes undertaken at the works and so in 1941 Waffen-SS units received a number of tank-hunter SP-chassis based on the Sd.Kfz.10 with 5-cm Pak 38 L/60 mounted on a pivot on the loading surface and minus wheels and chassis tail. The bonnet and driver-front area were lightly armoured. For 1941, the Pak 38 was a customer to be taken very seriously. It fired the Panzergranate 39 shell through 78 mm armour at a range of 500 metres. The gun was less effective against the T-34 and KW-1 on the Eastern Front.

Sd.Kfz.10 with 5-cm Pak 38.
(Vincent Bourguignon)

7.62-cm-Pak 36(r) auf 5 t Zugkraftwagen Diana Sd.Kfz. 6/3

Type:	Anti-tank gun on SP-chassis
Manufacturer:	Büssing-NAG, Alkett
Fighting weight:	11,200 kg
Length:	6330 mm
Breadth:	2260 mm
Height:	2980 mm
Motor:	Maybach HL 54 TURKM 6-cylinder petrol engine
Cubic capacity:	5420 ccm
Performance kW/hp:	73.6/115
Performance weight:	10.3 hp/t
Top speed	50 km/hr (road)
Fuel capacity:	190 litres
Range:	300 km (road) 115 km (terrain)
Crew:	6
Armament:	1 x 7.62 cm Pak 36 (r) L/51
Armour: max	10 mm
Fording depth:	0.6 m

At the beginning of 1942 the Altmärkische Kettenfabrik (Alkett) built two slightly differing versions of a tank-hunter SP-chassis, nine vehicles in all, based on the Büssing-NAG 5-tonne towing tractor. A high, open-top armoured superstructure with straight sides was fitted around the Sd.Kfz.6 loading surface. The motor and driver areas on the other hand remained without armour. The superstructure had large access doors either side and could be covered over by a tarpaulin attached to three loops. Armament waas the 7.62-cm Pak of Soviet origin. In the first months of the Russian campaign numerous type M 1936 field guns of 7.62 cm calibre had been captured. The Wehrmacht put these efficient weapons to use without (or with only slight) modifications as 7.62-cm FK 296(r) and 7.62-cm Pak 36(r). All nine were attached in 1942 to Panzerjäger-Abt.605 in North Africa. Although the firepower of all panzer-paks in this theatre of war was up to the mark, the "Diana" tank-hunters with their clumsy look and lack of armour failed to imprress.

A Diana captured by British forces in Libya. (IWM)

2-cm-Flakvierling (Sf) auf Zugkraftwagen 8t Sd-Kfz. 7/1

Type:	Quadruple flak gun on SP-chassis
Manufacturer:	Krauss-Maffei, Daimler-Benz, Büssing-NAG, Sauerer, Borgward
Fighting weight:	11,540 kg
Length:	6850 mm
Breadth:	2440 mm
Height:	3200 mm
Motor:	Maybach HL 62 TUK 6-cylinder petrol engine
Cubic capacity:	6191 ccm
Performance kW/hp:	103/140
Performance weight:	12.1 hp/t
Top speed:	50 km/hr (road)
Fuel capacity:	213 litres, later 203 litres
Range:	250 km (road) 135 km (terrain)
Crew:	9
Armament:	1 x 2 cm quadruple flak 38 L/65
Armour: max:	8 mm
Fording depth:	0.65 m

Besides its main purpose as a tractor, the Sd.Kfz.8 was also the basis for SP-flak guns. The first model of this kind was the Sd.Kfz.7/1 fitted with a 2-cm quadruple flak. These vehicles, initially without any armour, had a six-man crew. The flak mounting was set on the modified loading surface of the ZgKw and could by traversed by hand through 360°, elevation ranged from -10° to +100°. Theoretical rate of fire was 1800 rounds per minute. Besides 600 rounds ready ammunition another 1800 rounds and other equipment was carried in a single-axled trailer (Sd.Anh.56). In the course of the war the flak gun was given a protective shield; later the cooling system and driver space were also provided with light armour to reduce losses in personnel. The Sd.-Kfz.7/1 was also used against ground targets.

Sd.Kfz.7/1 of Panzergrenadier-Division Grossdeutschland, Eastern Front 1943. (Vincent Bourguignon)

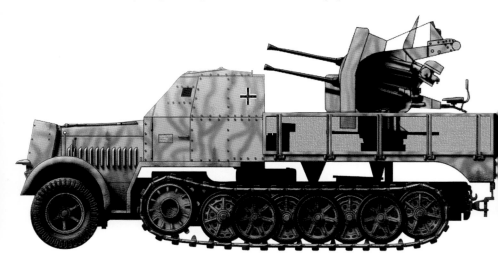

3.7-cm-Flak (Sf) auf Zugkraftwagen 8t Sd.Kfz.7/2

Type:	Self-propelled flak gun
Manufacturer:	Krauss-Maffei, Daimler-Benz, Büssing-NAG, Sauerer, Borgward
Fighting weight:	11,050 kg
Length:	6850 mm
Breadth:	2440 mm
Height:	3200 mm
Motor:	Maybach HL 62 TUK 6-cylinder petrol engine
Cubic capacity:	6191 ccm
Performance kW/hp:	103/140
Performance weight:	12.7 hp/t
Top speed:	50 km/hr (road)
Fuel capacity:	213 litres, later 203 litres
Range:	250 km (road), 135 km (terrain)
Crew:	7
Armament:	1 x 3.7 cm flak 36/37/43 L/98
Armour: max.	8 mm
Fording depth:	0.65 m

Besides the 2-cm flak quadruple the 3.7-cm flak was also installed on the 8-tonne ZgKw. Irrespective of which version of the 3.7-cm was fitted, the official designation was "3.7-cm Flak 36/37/43(Sf) auf ZgKw 8 t". The Flak 36 and 37 had a cadence of 140 to 160 rounds per minute, while the rate of fire of the Flak 43 was 230 to 250 rounds/min. Each of these weapons had a traverse of 360°: the Flak 36 and 37 had a range of elevation between -8° to +85°, the Flak 43 -7.5° to +90°. The crew was one driver, five gunners and a commander. The vehicle carried 120 rounds of flak ammunition, another 432 rounds and equipment were held in the single-axle trailer (Sd.Anh.57). Between 1942 and 1945 around 1000 vehcicles of this type were built. The first of the Sd.Kfz.7/2 had no armour, from 1943 a shield was provided for the gun and light armour for the driver's cabin and cooling system. As with the Sd.Kfz.7/1, during journeys the gunners sat against the rear bulkhead of the driver's cabin. The elongated sides and roof armour gave them a certain protection.

Sd.Kfz.7/2 in the Russian winter, beginning 1944. (WKA)

8.8-cm-Flak 18 (Sf) auf ZgKw 12t Sd.Kfz.8

Type:	Self-propelled anti-tank gun
Manufacturer:	Daimler-Benz
Fighting weight:	20 tonnes
Length:	7350 mm
Breadth:	2500 mm
Height:	Not available
Motor:	Maybach HL 85 TURKM 12-cylinder petrol engine
Cubic capacity:	8520 ccm
Performance kW/hp:	136/185
Performance weight:	9.25 hp/t
Top speed:	50 km/hr (road)
Fuel capacity:	250 litres
Range:	250 km (road) 100 km (terrain)
Crew:	9
Armament:	1 x 8.8 cm flak 18 L/56
Armour:	6 to 14.5 mm
Fording depth:	0.63 m

The 12-tonne tractor (Sd.Kfz.8) developed by Daimler-Benz was series produced as from 1934 and was used primarily by the Wehrmacht as a tractor for heavy artillery. By 1944 around 4,000 had left the various production belts in the work halls. In 1939 the Weapons Department ordered twelve SP-chassis built based on the Sd.Kfz.8 (bogie type DB 9). The vehicles were equipped on the loading surface with an 8.8-cm flak 18 protected by a large shield. The engine and driver's spaces were lightly armoured since the intention was basically to use the gun against tanks and field fortifications. The anti-tank SP guns of this type were used operationally in Poland and the French campaign. The disadvantage was their size and high profile which made it difficult to hide them.

On the basis of the 18-tonne FAMO tractor (Sd.Kfz.9), Weserhütte/Bad Oeynhausen built 14 similar examples fitted with an 8.8-cm Flak 37. In this case however there was no series production.

Sd.Kfz.8 with 8.8 cm flak 18(Sf), Western Front 1940. (WKA)

Leichter Wehrmachtschlepper (leWS)

Type:	Light armoured half-track tractor
Manufacturer:	Adler
Fighting weight:	6,900 kg
Length:	5200 mm
Breadth:	2120 mm
Height:	2 metres
Motor:	Maybach HL 30 6-cylinder petrol engine
Cubic capacity:	Not available
Performance kW/hp:	70/95
Performance weight:	13.8 hp/t
Top speed:	23 km/hr (road)
Fuel capacity, range, fording depth:	Details not available
Crew:	2
Armament:	Projected
Armour	6 to 15 mm

In May 1942 Adler of Frankfurt/Main began development work on a simple light tractor to replace the 1-tonne DEMAG model. Series production was set for the beginning of 1943 but the so-called light Wehrmacht tractor never left the experimental stage. Only three slightly different prototypes were completed. The first two had a 95-hp Maybach HL 30 engine, the third a 100-hp Maybach HL 42.

The vehicles had ungreased crawler tracks and a simplified wheel system better able to cope with mud and frost than the boxed tracks of the usual tractors. The driver's cabin and engine space were armoured since the intention was to arm the vehicle. For this reason support mounts were installed on the flat loading surface for various weapons.

Light Wehrmacht tug. (WKA)

15-cm-Panzerwerfer 42 (Sf) Sd.Kfz. 4/1

Type:	Self-propelled rocket launcher
Manufacturer:	Opel
Fighting weight:	8,500 kg
Length:	6 metres
Breadth:	2200 mm
Height:	3050 mm
Motor:	Opel 6-cylinder petrol engine
Cubic capacity:	3600 ccm
Performance kW/hp:	55/75
Performance	8.8 hp/t
Top speed:	40 km/hr (road)
Fuel capacity:	Not available
Range:	130 km (road) 80 km (terrain)
Crew:	4
Armament:	1 x 150 mm Panzerwerfer 42, 1 x 7.92 mm MG 34 or MG 42
Armour:	6 to 10 mm
Fording depth:	0.44 m

The appalling roads and by-ways on the Eastern Front threatened to make it impossible for Wehrmacht lorries to continue their role as transporters and supply vehicles. Relief was promised by simple half-track vehicles. Using crawler tracks to replace the rear wheels on normal 3-tonne and 4.5-tonne lorries of various manufacturers, the Wehrmacht intended to give these vehicles better cross-country capability. The crawler-track lorries of this kind were all known as "Maultiere" (mules). Between 1942 and 1945 around 22,000 were turned out. In 1943 Opel created on the basis of the 3-tonne Maultier from its own factories a lightly armoured SP-chassis for the Panzerwerfer 42. The 10-barrelled rocket launcher with 360° traverse was mounted on the rear superstructure. The 15-cm rockets had a range of up to 6,700 metres and each weighed 34.7 kg. The vehicle carried a stock of twenty. Waffen-SS units used on the same basis an 8-cm rocket launcher with 24 firing rails as fitted to the medium APC S303(f). A

7.92-mm MG was often located on the driver's cabin for protection. From April 1943 until March 1944 Opel produced 300 Sd.Kfz.4/1 and an additional 289 ammunition carriers Sd.Kfz.4 to support the rocket launchers. Later nineteen of these ammunition carriers were converted. There were three lots of the Sd.Kfz.4 and 4/1 built differing by changes to the superstructure and crawler tracks.

Sd.Kfz.4/1, unknown unit, eastern Front, beginning of 1945. (Vincent Bourguignon)

Above: View of an Sd.Kfz.4/1 with rear doors clearly visible. (US Army)

Below: A Panzerwerfer rocket launcher being loaded from within an Sd.Kfz.4/1. (NARA)

Schwerer Wehrmachtsschle pper (sWS) (gep. Ausf.)

Type:	Armoured half-track tug
Manufacturer:	Büssing-NAG, Tatra
Fighting weight:	13.5 tonnes
Length:	6675 mm
Breadth:	2500 mm
Height:	2830 mm
Motor:	Maybach HL 42 TRKM 6-cylinder petrol engine
Cubic capacity:	4170 ccm
Performance kW/hp:	73.6/100
Performance weight:	7.4 hp/t
Top speed:	29 km(hr (road)
Fuel capacity:	240 litres
Range:	300 km (road) 150 km (terrain)
Crew:	2
Armament:	See variants
Armour:	6 to 15 mm
Fording depth:	1 m

The heavy Wehrmacht Tractor (sWS) was conceived as a very simple, mass-produced and economic successor to the 3-tonne and 5-tonne tractors Sd.Kfz.11 and Sd.Kfz.6. In May 1942 the Army Weapons Department distributed development contracts, but between December 1943 and March 1945 Büssing-NAG and Tatra built only 825 (other sources speak of 1000) of these vehicles. The sWS

The unarmed armoured version of the sWS was intended to supply frontline troops. (US Army)

The flak crew continued to be exposed to enemy fire as before. (WKA)